Electrode Potentials

Richard G. Compton
Giles H. W. Sanders

Physical and Theoretical Chemistry Laboratory and St John's College, University of Oxford

Series sponsor: **ZENECA**

ZENECA is a major international company active in four main areas of business: Pharmaceuticals, Agrochemicals and Seeds, Specialty Chemicals, and Biological Products.

ZENECA's skill and innovative ideas in organic chemistry and bioscience create products and services which improve the world's health, nutrition, environment, and quality of life.

ZENECA is committed to the support of education in chemistry.

OXFORD NEW YORK TOKYO
OXFORD UNIVERSITY PRESS
1996

Oxford University Press, Walton Street, Oxford OX2 6DP

Oxford New York
Athens Auckland Bangkok Bombay
Calcutta Cape Town Dar es Salaam Delhi
Florence Hong Kong Istanbul Karachi
Kuala Lumpur Madras Madrid Melbourne
Mexico City Nairobi Paris Singapore
Taipei Tokyo Toronto
and associated companies in
Berlin Ibadan

Oxford is a trade mark of Oxford University Press

Published in the United States
by Oxford University Press Inc., New York

A catalogue record for this book is available from the British Library

Library of Congress Cataloging in Publication Data
Compton, R. G.
Electrode potentials / Richard G. Compton, Giles H. W. Sanders.
(Oxford chemistry primers; 41)
Includes index.
1. Electrodes. 2. Electrochemistry I. Sanders, Giles H. W.
II. Title. III. Series.
QD571.C65 1996 541.3'724–dc20 95-52660

ISBN 0 19 8556845

Typeset by EXPO Holdings, Malaysia
Printed in Great Britain by
The Bath Press, Somerset

Founding Editor's Foreword

Electrode potentials is an essential topic in all modern undergraduate chemistry courses and provides an elegant and ready means for the deduction of a wealth of thermodynamic and other solution chemistry data. This primer develops the foundations and applications of electrode potentials from first principles using a minimum of mathematics only assuming a basic knowledge of elementary thermodynamics.

This primer therefore provides an easily understood and student-friendly account of this important topic and will be of interest to all apprentice chemists and their masters.

Stephen G. Davies
The Dyson Perrins Laboratory
University of Oxford

Preface

This Primer seeks to provide an introduction to the science of equilibrium electrochemistry; specifically it addresses the topic of electrode potentials and their applications. It builds on a knowledge of elementary thermodynamics giving the reader an appreciation of the origin of electrode potentials and shows how these are used to deduce a wealth of chemically important information and data such as equilibrium constants, the free energy, enthalpy and entropy changes of chemical reactions, activity coefficients, the selective sensing of ions, and so on. The emphasis throughout is on understanding the foundations of the subject and how it may be used to study problems of chemical interest. The primer is directed towards students in the early years of their university courses in chemistry and allied subjects; accordingly the mathematical aspects of the subject have been minimised as far as is consistent with clarity.

We thank John Freeman for his skilful drawing of the figures in this primer. His patience and artistic talents are hugely appreciated.

Oxford R. G. C. and G. H. W. S.
September 1995

Contents

1 Getting started

1.1 The scope and nature of this primer

The aim of this primer is to provide the reader with a self-contained, introductory account of the science of electrochemistry. It seeks to explain the origin of electrode potentials, show their link with chemical thermodynamics and to indicate why their measurement is important in chemistry. In so doing some ideas about solution non-ideality and how ions move in solution are helpful, and essential diversions into these topics are made in Chapters 2 and 3.

1.2 The origin of electrode potentials

Figure 1.1 shows the simplest possible electrochemical experiment. A metal wire, for example made of platinum, has been dipped into a beaker of water which also contains some Fe(II) and Fe(III) ions. As the aqueous solution will have been made by dissolving salts such as $Fe(NO_3)_2$ and $Fe(NO_3)_3$ there will inevitably be an anion, for example NO_3^-, also present. This anion is represented by X^- and since we expect the solution to be uncharged ('electroneutral'),

$$[X^-] = 2[Fe^{2+}] + 3[Fe^{3+}]$$

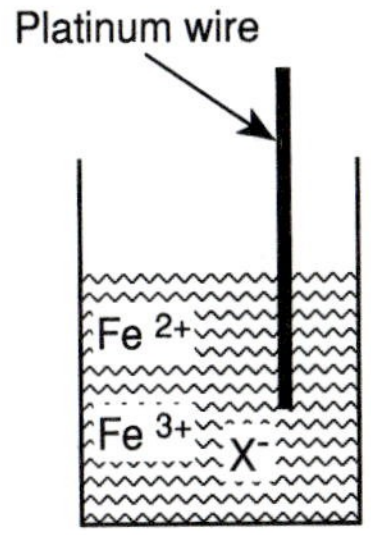

Fig. 1.1 A metal wire in a solution containing Fe(II) and Fe(III) ions.

Considering the relative electronic structures of the two cations in the solution we note that the two metal ions differ only in that Fe(II) contains one extra electron. It follows that the ions may be interconverted by adding an electron to Fe(III) ('reduction') or by removing an electron from Fe(II) ('oxidation').

In the experiment shown in the figure the metal wire can act as a source or sink of a tiny number of electrons. An electron might leave the wire and join an Fe^{3+} ion in the solution, so forming an Fe^{2+} ion. Alternatively an Fe^{2+} cation close to the electrode might give up its electron to the metal so turning itself into an Fe^{3+} ion. In practice both these events take place and very shortly after the wire ('electrode') is placed in the solution the following *equilibrium* is established at the surface of the metal:

$$Fe^{3+}(aq) + e^-(metal) \rightleftharpoons Fe^{2+}(aq) \qquad (1.1)$$

The equilibrium symbol, $\rightleftharpoons$, has the same meaning here as when applied to an ordinary chemical reaction and indicates that the forward reaction {here $Fe^{3+}(aq) + e^-(metal) \rightarrow Fe^{2+}(aq)$} and the reverse reaction {$Fe^{2+}(aq) \rightarrow Fe^{3+}(aq) + e^-(metal)$} are *both* occurring and are taking place at the same rate so that there is no further *net* change.

Equation (1.1) merits further reflection. Notice the forward and reverse processes involve the transfer of electrons between the metal and the solution phases. As a result when equilibrium is attained there is likely

A phase is a state of matter that is uniform throughout, both in chemical composition and in physical state. Thus ice, water and steam are three separate phases as are diamond, graphite and C_{60}.

to be a net electrical charge on each of these phases. If the equilibrium shown in eqn (1.1) lies to the left in favour of the species Fe^{3+}(aq) and e^-(metal), then the electrode will bear a net negative charge and the solution a net positive charge. Conversely if the equilibrium favours Fe^{2+}(aq) and lies to the right, then the electrode will be positive and the solution negative. Regardless of the favoured direction, it can be expected that at equilibrium there will exist a charge separation and hence a potential difference between the metal and the solution. In other words an *electrode potential* has been established on the metal wire relative to the solution phase. The chemical process given in eqn (1.1) is the basis of this electrode potential: throughout the rest of this primer we refer to the chemical processes which establish electrode potentials, as *potential determining equilibria*. Equation (1.1) describes the potential determining equilibrium for the system shown in Fig. 1.1.

It has been suggested that rattlesnakes shake their rattles to charge themselves with static electricity (*Nature*, 370, 1994, p. 184). This helps them locate sources of moist air in the environment since plumes of such air, whether from a sheltered hole or an exhaling animal, pick up electric charge from the ground and may be detectable by the tongue of the charged snake as it moves back and forth. Experiments in which a rattle (without its former owner) was vibrated at 60 Hz produced a voltage of around 75–100 V between the rattle and earth by charging the former.

Box 1.1 Electrical potential

Suppose the rattlesnake shown below may be either positively or negatively charged. To examine the sign and magnitude of the rattlesnake's charge, the following 'thought experiment' can be envisaged. A test probe bearing a unit positive charge of one coulomb is moved from a point infinitely distant from the rattlesnake and brought up to it. The *work* required to do this is measured and is the *electrical potential*, ϕ, of the rattlesnake.

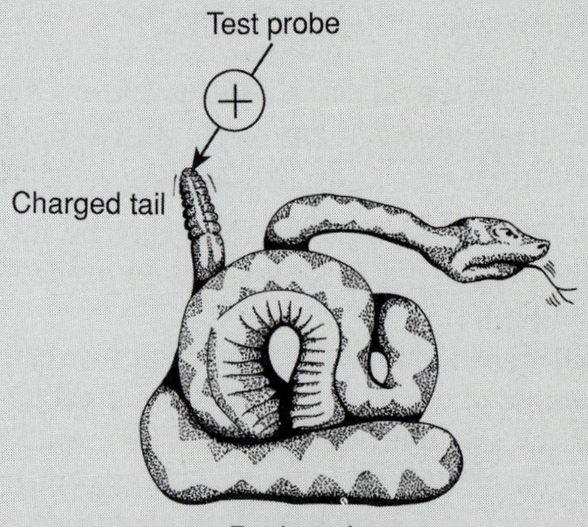

An imaginary experiment is depicted in the box: a probe carrying one coulomb of positive charge is moved from an infinitely distant point to the charged tail of a rattlesnake.

The diagram shows that if the rattlesnake is positively charged then the work in moving the probe will be positive and the rattlesnake will bear a positive electrical potential. Conversely if the charge on the rattlesnake is negative then the work of transfer is negative—energy is released—and the rattlesnake bears a negative electrical potential.

In the main text the electrical potential of a metal wire or electrode, ϕ_M, and of a solution, $\phi_{solution}$, are discussed. The absolute measurement of these quantities might, in principle, proceed as illustrated above for the rattlesnake, albeit, possibly with less excitement.

The ions Fe^{2+} and Fe^{3+} feature in the potential determining equilibrium given in eqn (1.1). It may therefore be correctly anticipated that the magnitude and sign of the potential difference on the platinum wire in Fig. 1.1 will be governed by the relative amounts of Fe^{2+} and Fe^{3+} in the solution.

To explore this dependence consider what happens when a further amount of $Fe(NO_3)_3$ is added to the solution thus perturbing the equilibrium:

$$Fe^{3+}(aq) + e^{-}(metal) \rightleftharpoons Fe^{2+}(aq).$$

This will become 'pushed' to the right and electrons will be removed from the metal. Consequently the electrode will become more positive relative to the solution. Conversely addition of extra $Fe(NO_3)_2$ will shift the equilibrium to the left and electrons will be added to the electrode. The latter thus becomes more negative relative to the solution.

In considering shifts in potential induced by changes in the concentrations of Fe^{3+} or Fe^{2+} it should be recognised that the quantities of electrons exchanged between the solution and the electrode are infinitesimally small and too tiny to directly measure experimentally.

We have predicted that the potential difference between the wire and the solution will depend on the amount of Fe^{3+} and Fe^{2+} in solution. In fact it is the ratio of these two concentrations that is crucially important. The potential difference is given by

$$\phi_M - \phi_{solution} = \text{constant} - \frac{RT}{F}\ln\left\{\frac{[Fe^{2+}]}{[Fe^{3+}]}\right\} \qquad (1.2)$$

where ϕ denotes the electrical potential. ϕ_M is the potential of the metal wire (electrode) and $\phi_{solution}$ the potential of the solution phase. Equation (1.2) is the famous Nernst equation. It is written here in a form appropriate to a single electrode/solution interface. Later in this chapter we will see a second form which applies to an electrochemical cell with two electrodes and hence two electrode/solution interfaces. The other quantities appearing in equation (1.2) are

R = the gas constant (8.313 J K^{-1} mol^{-1})
T = absolute temperature (measured in K)
F = the Faraday constant (96487 C mol^{-1})

As emphasised above, when equilibrium (1.1) is established, this involves the transfer of an infinitesimal quantity of charge and hence the interconversion of only a vanishingly small fraction of ions. Consequently the concentrations of Fe(II) and Fe(III) in eqn (1.2) are imperceptibly different from in those in the solution before the electrode (wire) was inserted into it.

The shift in electrode charge resulting from the addition of Fe^{2+} or Fe^{3+} may be thought of as an extension of Le Chatelier's Principle which is often used as a guide to the prediction of temperature, pressure and other effects on chemical equilibria. The principle is applied as follows:- *Suppose a change (of temperature, pressure, chemical composition, ...) is imposed on a system previously at equilibrium. Le Chatelier's Principle predicts that the system will respond in a way so as to oppose or counteract the imposed perturbation.* For example:-

- an increase in pressure shifts the equilibrium $N_2(g) + 3H_2(g) \rightleftharpoons 2NH_3(g)$ more in favour of NH_3 since the reaction proceeds with a net loss of molecules. This reduction in the total number of molecules will tend to oppose the applied increase in pressure.
- an increase in temperature shifts the equilibrium NH_4NO_3 (s) + H_2O(l) $\rightleftharpoons$ NH_4^+ (aq) + NO_3^- (aq) more in favour of the dissolved ions since the dissolution is an endothermic process. This loss of enthalpy will tend to oppose the applied increase in temperature.
- an increase in $[Fe^{3+}]$ shifts the equilibrium e^- (metal) + Fe^{3+} (aq) $\rightleftharpoons$ Fe^{2+} (aq) more in favour of the Fe^{2+} ion. This reduces the imposed increase in $[Fe^{3+}]$ and makes the metal more positively charged.

The Faraday constant represents the electrical charge on one mole of electrons so $F = e.N_A$ where e is the charge on a single electron and N_A is the Avogadro Number. e has the value 1.602×10^{-19} C and N_A the value 6.022×10^{23} mol^{-1}.

1.3 Electron transfer at the electrode/solution interface

We now consider further the experiment introduced in the previous section. It is helpful to focus on the energy of electrons in the metal wire and in the Fe^{2+} ions in solution as depicted in Fig. 1.2. Note that in the figure an empty level on Fe^{3+} is shown. This corresponds to an unfilled d orbital. When this orbital gains an electron the metal ion is reduced and becomes Fe^{2+}. The electronic structure of a metal is commonly described by the 'electron sea' model in which the conduction electrons are free to

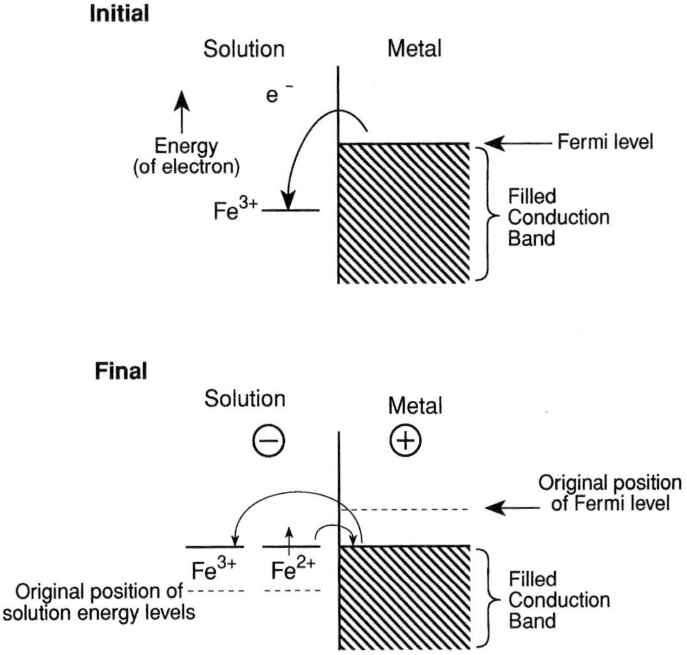

Fig. 1.2 The energy of electrons in ions in solution and in the metal wire depicted in Fig. 1.1.

move throughout the solid binding the cations rigidly together. Energetically the electrons form into 'bands' in which an effective continuum of energy levels are available. These are filled up to a energy maximum known as the Fermi level. In contrast electrons located in the two solution phase ions—Fe^{2+} and Fe^{3+}— are localised and restricted to certain discrete energy levels as implied in Fig. 1.2. The lowest empty level in Fe^{3+} is close in energy to the highest occupied level in Fe^{2+} as shown. Note however these levels do not have *exactly* the same energy value, since adding an electron Fe^{3+} will alter the solvation around the ion as it changes from Fe^{3+} to Fe^{2+}. The upper part of Fig. 1.2 shows the position of the Fermi level relative to the ionic levels the very instant that the metal is inserted into the solution and before *any* transfer of electrons between the metal and the solution has occurred. Notice that as the Fermi level lies above the empty level in Fe^{3+} it is energetically favourable for electrons to leave the metal and enter the empty ionic level. This energy difference is the 'driving force' for the electron transfer we identified as characteristic of the experiment shown in Fig. 1.1.

What is the consequence of electrons moving from the metal into the solution phase? The metal will become positively charged while the solution must become negative: *this charge transfer is the fundamental reason for the potential difference predicted by the Nernst equation.* In addition as electron transfer proceeds, and the solution and metal become charged, the energy level both in the metal and in solution must change. Remember that the vertical axis in Fig. 1.2 represents the energy of an electron. Thus if positive charge evolves on the electrode then the energy of an electron in the metal must be lowered, and so the Fermi level must lie progressively further down the diagram. This is illustrated in the lower

When ions such as iron(II) or iron(III) exist in water they are *hydrated.* That is a number of water molecules—probably six in these cases—are relatively tightly bound to the ion. This serves to stabilise the ion and is an important driving force which encourages the dissolution of solids such as $Fe(NO_3)_3$ and $Fe(NO_3)_2$ in water. The highly charged ions may also orientate or partially orientate more distant water molecules. The water molecules directly attached to the ion comprise its inner or primary hydration shell and the other solvent molecules perturbed by the ion constitute an outer hydration shell.

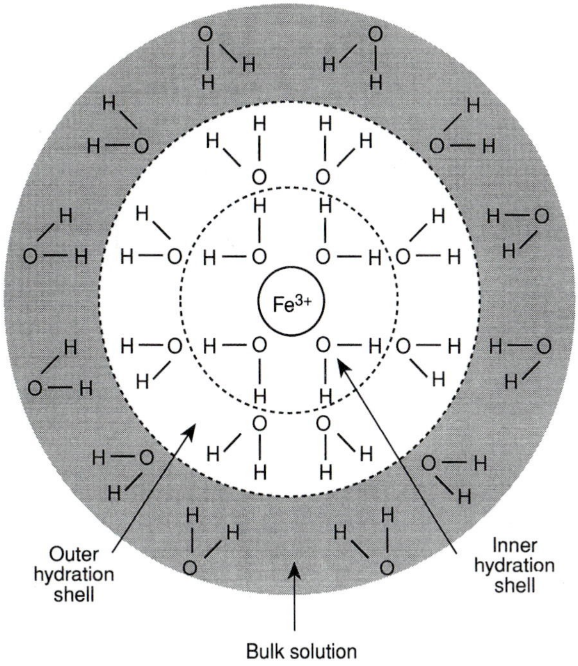

A schematic diagram of the hydration of a Fe^{3+} ion showing the inner and outer hydration shells.

part of the picture. Equally the generation of negative charge on the solution must destabilise the electron energies within ions in that phase and the energy levels describing Fe^{3+} and Fe^{2+} will move upwards.

We can now see why it is that the electron transfer between metal and solution rapidly ceases before significant measurable charge can be exchanged. This is because the effect of charge transfer is to move the ionic levels and the Fermi level towards each other and hence reduce, and ultimately destroy, the driving force for further electron transfer.

The pictorial model outlined leads us to expect that when the metal and solution are at equilibrium this will correspond to an exact matching of the energy levels in the solution with the Fermi level. When this point is reached there will be a difference of charge and hence of potential between the metal and solution phases. This is the basic origin of the Nernst equation outlined earlier and which we will shortly derive in more general terms once we have briefly reviewed how equilibrium is described by the science of chemical thermodynamics.

1.4 Thermodynamic description of equilibrium

Consider the following gas phase reaction

$$A(g) \rightleftharpoons B(g) \tag{1.3}$$

The simplest way of keeping track of this system is to note that at equilibrium the reactants and products must have identical chemical potentials so that,

$$\mu_A = \mu_B \tag{1.4}$$

Box 1.2 Chemical potentials

Thermodynamics tells us that the Gibbs free energy of a system, G_{sys}, is minimised when it has attained equilibrium. This is illustrated below:

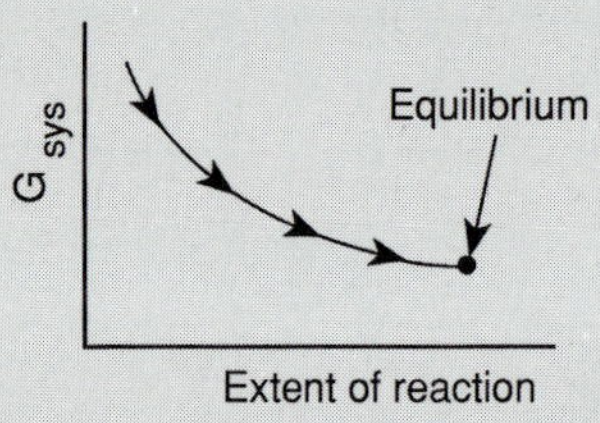

As a system reaches equilibrium the Gibbs free energy decreases, attaining a minimum at the point of equilibrium.

Mathematically this minimisation is expressed as

$$dG_{sys} = 0 \text{ (at equilibrium)}$$

Consider the A/B isomerisation reaction discussed in the text. If there are n_A moles of A and n_B moles of B then dn_A represents the change in the number of moles of A and dn_B the corresponding quantity for B. However, since every A-molecule lost results in one molecule B formed,

$$-dn_A = +dn_B = dn \text{ (for convenience)}$$

The associated change in G_{sys} is given by

$$\begin{aligned} dG_{sys} &= \text{Free Energy Change on } \textit{formation} \text{ of B} + \text{Free Energy Change on } \textit{loss} \text{ of A} \\ &= \{\text{rate of change of } G_{sys} \text{ with } n_B\}dn_B \\ &\quad + \{\text{rate of change of } G_{sys} \text{ with } n_A\}dn_A \\ &= (\partial G/\partial n_B)dn_B + (\partial G/\partial n_A)dn_A \\ &= [(\partial G/\partial n_B) - (\partial G/\partial n_A)]dn \end{aligned}$$

The partial derivatives in the square brackets are the chemical potentials of B and A respectively:

$$(\partial G/\partial n_B) = \mu_B \quad \text{and} \quad (\partial G/\partial n_A) = \mu_A$$

so that

$$dG_{sys} = [\mu_B - \mu_A]dn$$

It follows that at equilibrium,

$$\mu_A = \mu_B$$

that is, the chemical potential of the reactant(s) equals that of the product(s).

This is an alternative but more convenient form of the more familiar statement that at equilibrium the total Gibbs free energy of a system will be a minimum. (If you are unfamiliar with the former approach Box 1.2 may convince you of its equivalence to the latter.) We can relate chemical potentials to the *partial pressures*, P_A and P_B, of the gases concerned if we assume them to be ideal:

$$\mu_A = \mu_A^o + RT \ln P_A \text{ and } \mu_B = \mu_B^o + RT \ln P_B \qquad (1.5)$$

where $\mu_A{}^o$ and $\mu_B{}^o$ are the standard chemical potentials of the gases A and B and have a constant value (at a fixed temperature). It follows that at equilibrium,

$$K_p = P_B/P_A = \exp\{(\mu_A^o - \mu_B^o)/RT\} \qquad (1.6)$$

This equation tells us that the ratio P_B/P_A is fixed and constant. K_p will be familiar to you as the *equilibrium constant* for the reaction.

Let us remind ourselves what happens if we consider the same reaction as before but now suppose that it is carried out in solution,

$$A(aq) \rightleftharpoons B(aq) \qquad (1.7)$$

We can again apply eqn (1.4) but now have a choice as to whether we use mole fractions (x) or concentrations ([ ]):

$$\mu_A = \mu_A^\nabla + RT \ln x_A \text{ and } \mu_B = \mu_B^\nabla + RT \ln x_B \qquad (1.8a)$$

or

$$\mu_A = \mu_A^\ominus + RT \ln [A] \text{ and } \mu_B = \mu_B^\ominus + RT \ln [B] \qquad (1.8b)$$

where the solutions are assumed to be ideal. If there are n_A moles of A and n_B moles of B in the solution of volume V, then

$$x_A = n_A/(n_A + n_B) \text{ and } x_B \; n_B/(n_A + n_B)$$

and

$$[A] = n_A/V \text{ and } [B] = n_B/V.$$

The choice leads to two alternative standard states:

(i) when mole fractions are used (as in eqn 1.8a) μ^∇ is the chemical potential when $x = 1$ and so relates to a pure liquid, and

(ii) when considering concentrations (as in eqn 1.8b) $\mu^\ominus$ is the chemical potential of a solution of A of unit concentration, $[A] = 1$ mol. dm^{-3}.

Two types of equilibrium constant result:

$$K_x = x_B/x_A \qquad (1.9a)$$

and

$$K_c = [B]/[A] \qquad (1.9b)$$

Provided, as we have supposed, the solutions are ideal the two descriptions of equilibrium are the same. This follows since

$$K_x = \frac{n_A/(n_A + n_B)}{n_B/(n_A + n_B)} = \frac{n_A}{n_B} = \frac{n_A/V}{n_B/V} = K_C \qquad (1.10)$$

Chemical potentials are simply related to the Gibbs free energy of the system, G, through the equations $\mu_A = (\partial G/\partial n_A)$ and $\mu_B = (\partial G/\partial n_B)$ where n_A and n_B are the number of moles of A and B respectively. In a system comprising only pure A then μ_A is simply the Gibbs free energy of one mole of A.

The sum of the partial pressures equals the total pressure P_{TOTAL}. In this example $P_A + P_B = P_{TOTAL}$

The standard chemical potential, μ^o, here is the Gibbs free energy of one mole of the pure gas at one atmosphere pressure. This follows from substituting $P = 1$ into equation (1.5) combined with the definitions of μ_A and μ_B.

The standard chemical potentials μ^∇, $\mu^\ominus$ and μ^o depend on the temperature chosen to define the standard state.

1.5 Thermodynamic description of electrochemical equilibrium

Both the equilibria considered in Section 1.4 occur in a single phase—either in the gas phase or in solution. However we saw earlier that the essential feature of an electrochemical equilibrium, such as the one displayed in Fig. 1.1, is that it involves two separate phases, the electrode and the solution. Moreover the equilibrium involves the transfer of a charged particle, the electron, between these two phases. This complicates the approach we have adopted since we have now to concern ourselves not only with differences in the *chemical energy* of the reactants and products (as in eqn 1.4) but also with *electrical energy* differences. The latter arise since typically a difference in potential exists between the solution and the metal electrode so that the relative electrical energy of the electron in the two phases helps control the final point of equilibrium.

We introduce a new quantity, the electrochemical potential, $\bar{\mu}_A$, of a species A

$$\bar{\mu}_A = \mu_A + z_A F\phi \tag{1.11}$$

where z_A is the charge on the molecule A. The electrochemical potential of A is thus comprised of two terms. The first is its chemical potential, μ_A. The second is a term, $z_A F\phi$, which describes the electrical energy of A. The form of this latter quantity is (charge multiplied by potential) which corresponds to an energy term; the factor F is required to put it on a per mole basis for use alongside μ_A which is likewise defined on a molar basis. The potential ϕ relates to the particular phase—electrode or solution—in which species A resides.

With this extension recognised we can now treat electrochemical equilibria in a manner analogous to our approach to the more familiar problems encountered in Section 1.4. By way of illustration let us return to the example of Section 1.2,

$$Fe^{3+}(aq) + e^{-}(metal) \rightleftharpoons Fe^{2+}(aq)$$

and derive the Nernst equation for this system. The starting point for this, and all subsequent examples, is that at equilibrium,

Total electrochemical potential of reactants
= Total electrochemical potential of products.

We apply eqn (1.11) to obtain,

$$(\mu_{Fe^{3+}} + 3F\phi_S) + (\mu_{e^-} - F\phi_M) = (\mu_{Fe^{2+}} + 2F\phi_S) \tag{1.12}$$

where ϕ_M and ϕ_S refer to the electrical potential of the metal electrode and of the solution respectively. The first term in round brackets refers to an Fe^{3+} ion in solution, the second to an electron in the metal and the third to an Fe^{2+} ion in solution. Rearranging,

$$F(\phi_M - \phi_S) = \mu_{Fe^{3+}} + \mu_{e^-} - \mu_{Fe^{2+}} \tag{1.13}$$

But,

$$\begin{aligned} \mu_{Fe^{3+}} &= \mu^{\ominus}_{Fe^{3+}} + RT\ln[Fe^{3+}] \\ \mu_{Fe^{2+}} &= \mu^{\ominus}_{Fe^{2+}} + RT\ln[Fe^{2+}] \end{aligned} \tag{1.14}$$

and hence

$$\phi_M - \phi_S = \Delta\phi^\ominus - \frac{RT}{F}\ln\left\{\frac{[Fe^{2+}]}{[Fe^{3+}]}\right\} \qquad (1.15)$$

which looks, finally, like the Nernst equation for the Fe(III)/Fe(II) system that we first cited in Section 1.2, provided the term

$$\Delta\phi^\ominus = \frac{1}{F}(\mu^\ominus_{Fe^{3+}} + \mu_{e^-} + \mu^\ominus_{Fe^{2+}}) \qquad (1.16)$$

is a constant. It is, since it contains two standard chemical potentials and a term, μ_{e^-} which is the chemical potential of an electron in the electrode.

1.6 Electrochemical experiments

The Nernst equation (eqn 1.15) just deduced for the Fe(III)/Fe(II) couple suggests two quite different types of electrochemical experiment:-

(i) Those in which an electrode dips into solution so that a potential is established on it in accordance with the predictions of the Nernst equation. We shall see that such measurements can, certainly in principle and often in practice, simply and conveniently yield precise sets of thermodynamic data (equilibrium constants, reaction free energies, enthalpies and entropies, and so on). This is the field of Nernstian or equilibrium electrochemistry. As the second name implies, no sustained currents flow during such experiments.

(ii) Those in which a potential is applied between the electrode and the solution, thus causing the concentrations of the species in the cell to adjust themselves so as to conform to the Nernst equation. In order for this to happen current has to flow and electrolysis takes place. This is the field of kinetic or dynamic electrochemistry and has synthetic importance and mechanistic interest.

Electrolysis reactions of industrial significance include the oxidation of brine to form chlorine, the winning of aluminium metal from the reduction of its molten ores and the reduction of acrylonitrile to form an important intermediate in the manufacture of nylon 66.

This primer is concerned almost exclusively with equilibrium electrochemistry.

A separate volume in the Oxford Chemistry Primers series is available to provide an introduction to kinetic electrochemistry (OCP 34, *Electrode Dynamics* by A. C. Fisher).

1.7 The Nernst equation and some other electrode/solution interfaces

We have seen how the concept of electrochemical potential has allowed us to develop the Nernst equation for the Fe(III)/Fe(II) system. In this

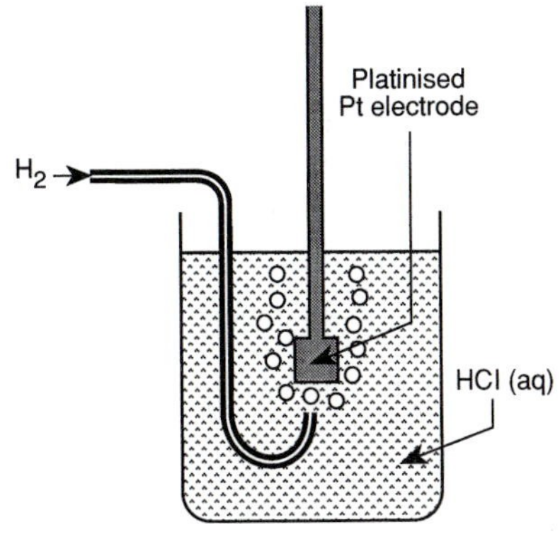

Fig. 1.3 A hydrogen electrode.

section we apply the same approach to three further systems before making some generalisations.

The hydrogen electrode

The electrode is formed by taking a platinum 'flag' electrode and electroplating a fine deposit of 'platinum black' from a solution containing a soluble platinum compound.

The first new system is shown in Fig. 1.3 and is the so-called hydrogen electrode. It comprises a platinum black electrode dipping into a solution of hydrochloric acid. Hydrogen gas is bubbled over the surface of the electrode. The reaction which determines the electrode potential again depends on the transfer of an electron between the Fermi level of the electrode and an ion in solution:

$$H^+(aq) + e^-(metal) \rightleftharpoons \tfrac{1}{2}H_2(g) \tag{1.17}$$

In this case the other participating species are located in the solution and in the gas phase. In order to predict the Nernst equation we again start with the idea that the total electrochemical potential of the reactants must equal the total of that of the products when equilibrium is established,

Total electrochemical potential of reactants
= Total electrochemical potential of products

Using the definition of electrochemical potential given in eqn (1.11) we find that

$$(\mu_{H^+} + F\phi_S) + (\mu_{e^-} - F\phi_M) = \tfrac{1}{2}\mu_{H_2} \tag{1.18}$$

It is excellent training for the reader to pick up the habit of checking that derived Nernst equations give the correct prediction for the change of electrode potential as positive or negative when tested against an alteration of the potential determining equilibrium for the electrode reaction in the light of Le Chatelier's Principle as illustrated in the text for the H^+/H_2 reaction. Benefits will accrue when it is required to solve problems later in this book.

where the first term in round brackets relates to the electrochemical potential of an H^+ ion in solution, the second to that of an electron in the platinum black and the third to gaseous H_2. The hydrogen term contains only a chemical potential but no electrical energy term since the molecule is uncharged. Rearrangement, together with the equations

$$\mu_{H^+} = \mu^{\ominus}_{H^+} + RT\ln[H^+] \tag{1.19}$$

and

$$\mu_{H_2} = \mu^{\circ}_{H_2} + RT\ln P_{H_2} \tag{1.20}$$

again gives us the desired Nernst equation:

$$\phi_M - \phi_S = \Delta\phi^{\ominus} + \frac{RT}{F}\ln\left\{\frac{[H^+]}{P^{1/2}_{H_2}}\right\} \tag{1.21}$$

where the constant

$$\Delta\phi^{\ominus} = \frac{1}{F}(\mu^{\ominus}_{H^+} + \mu_{e^-} - \tfrac{1}{2}\mu^{\circ}_{H_2}).$$

Notice the Nernst equation predicts that increasing $[H^+]$ should make the electrode more positive relative to the solution. This is exactly what we would predict on the basis of the potential determining equilibrium written for this electrode in eqn 1.17.

$$H^+(aq) + e^-(metal) \tfrac{1}{2} \rightleftharpoons H_2(g)$$

since, applying Le Chatelier's Principle, we would expect the equilibrium to be shifted to the right by added H^+. This would remove electrons from the electrode so making it more positively charged.

The chlorine electrode

We next turn to consider the chlorine electrode illustrated in Fig. 1.4. This comprises a bright platinum electrode in a solution containing chloride ions. Chlorine gas is bubbled over the electrode surface. The potential determining equilibrium is

$$\tfrac{1}{2}\mathrm{Cl_2(g)} + \mathrm{e^-(metal)} \rightleftharpoons \mathrm{Cl^-(aq)} \tag{1.22}$$

Using the definition of electrochemical potential given in eqn (1.11) we obtain

$$\tfrac{1}{2}\mu_{\mathrm{Cl_2}} + (\mu_{\mathrm{e^-}} - F\phi_{\mathrm{M}}) = (\mu_{\mathrm{Cl^-}} - F\phi_{\mathrm{S}}) \tag{1.23}$$

Rearrangement, whilst noting

$$\mu_{\mathrm{Cl_2}} = \mu^{\circ}_{\mathrm{Cl_2}} + RT\ln P_{\mathrm{Cl_2}} \tag{1.24}$$

and

$$\mu_{\mathrm{Cl^-}} = \mu^{\ominus}_{\mathrm{Cl^-}} + RT\ln[\mathrm{Cl^-}] \tag{1.25}$$

gives the relevant Nernst Equation for the chlorine electrode:

$$\phi_{\mathrm{M}} - \phi_{\mathrm{S}} = \Delta\phi^{\ominus} + \frac{RT}{F}\ln\left\{\frac{P^{1/2}_{\mathrm{Cl_2}}}{[\mathrm{Cl^-}]}\right\} \tag{1.26}$$

where the constant,

$$\Delta\phi^{\ominus} = \frac{1}{F}\left(\tfrac{1}{2}\mu^{\circ}_{\mathrm{Cl_2}} + \mu_{\mathrm{e^-}} - \mu^{\ominus}_{\mathrm{Cl^-}}\right).$$

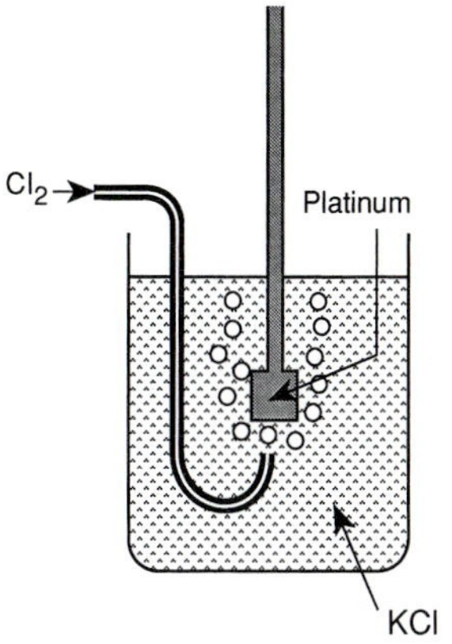

Fig. 1.4 A chlorine electrode.

Notice that equation (1.26) predicts that the quantity $\phi_M - \phi_S$ becomes more positive if the partial pressure of chlorine gas, p_{Cl_2}, is *increased* or if the chloride concentration, $[Cl^-]$ is *decreased.*

The silver/silver chloride electrode

As a final example, but one which introduces several new ideas, we consider the silver/silver chloride electrode shown in Fig. 1.5. A silver wire is coated with (porous) silver chloride by electro-oxidising the wire in a medium containing chloride ions such as an aqueous solution of KCl. The coated wire is then used in a fresh KCl solution as shown in Fig. 1.5. The following potential determining equilibrium establishes a potential on the silver electrode

'Electro-oxidising' means that electrolysis is used to bring about the reaction

$Ag(s) + Cl^-(aq) - e^-(metal) \rightarrow AgCl\ (s)$

$$\mathrm{AgCl(s)} + \mathrm{e^-(metal)} \rightleftharpoons \mathrm{Ag(metal)} + \mathrm{Cl^-(aq)} \tag{1.27}$$

The equilibrium is established at the silver/silver chloride boundary. It is therefore important that the silver chloride coat is *porous* so that the aqueous solution containing the chloride ions penetrates to the boundary and so permits the equilibrium in eqn (1.27) to be established.

Equating the electrochemical potentials of the reactants and products (in eqn 1.27) we obtain

$$\bar{\mu}_{\mathrm{AgCl}} + \bar{\mu}_{\mathrm{e^-}} = \bar{\mu}_{\mathrm{Ag}} + \bar{\mu}_{\mathrm{Cl^-}} \tag{1.28}$$

from which it follows, by using the definition of electrochemical potential, that

$$(\mu_{\mathrm{AgCl}}) + (\mu_{\mathrm{e^-}} - F\phi_{\mathrm{M}}) = (\mu_{\mathrm{Ag}}) + (\mu_{\mathrm{Cl^-}} - F\phi_{\mathrm{S}}) \tag{1.29}$$

Following our now familiar protocol we next expand the chemical potential terms of the chloride ion as follows

$$(\mu_{\mathrm{AgCl}}) + (\mu_{\mathrm{e^-}} - F\phi_{\mathrm{M}}) = (\mu_{\mathrm{Ag}}) + (\mu^{\ominus}_{\mathrm{Cl^-}} + RT\ln[\mathrm{Cl^-}] - F\phi_{\mathrm{S}}) \tag{1.30}$$

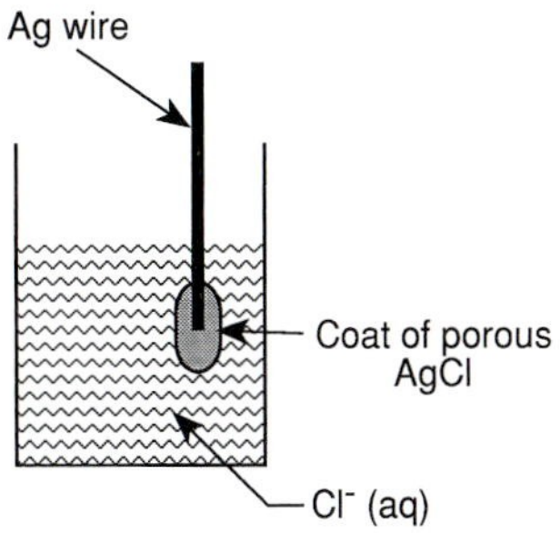

Fig. 1.5 A silver/silver chloride electrode.

However both AgCl and Ag are present as pure solids so

$$\mu_{AgCl} = \mu_{AgCl}^{\square} \tag{1.31}$$

$\mu_{AgCl}^{\square}$ and $\mu_{Ag}^{\square}$ are constant at a specified temperature.

and

$$\mu_{Ag} = \mu_{Ag}^{\square} \tag{1.32}$$

Notice that *no* terms of the form $RT\ln$[AgCl] or $RT\ln$[Ag] appear since these species are pure solids of fixed and definite composition. Such concentration terms only appear for solution phase species or for gases (where pressures replace [ ]) since the chemical potentials of these are given via equations such as

$$\mu_A = \mu_A^{o} + RT\ln P_A \text{ and } \mu_B = \mu_B^{o} + RT\ln P_B$$

or,

$$\mu_A = \mu_A^{\ominus} + RT\ln[A] \text{ and } \mu_B = \mu_B^{\ominus} + RT\ln[B].$$

In the case of pure solids however,

$$\mu_A = \mu_A^{\square}$$

Returning to eqn (1.30) and including eqns (1.31) and (1.32) gives

$$\phi_M - \phi_S = \Delta\phi^{\ominus} - \frac{RT}{F}\ln[Cl^-] \tag{1.33}$$

where

$$\Delta\phi^{\ominus} = \mu_{Ag}^{\square} + \mu_{Cl^-}^{\ominus} - \mu_{AgCl^-}^{\square} - \mu_{e^-} \tag{1.34}$$

Equation 1.30 is the Nernst equation for the silver/silver chloride electrode.

1.8 Concentrations or activities?

In the above we have assumed all the solutions phase species, such as H^+ or Cl^-, behave 'ideally'. Accordingly concentrations were used to relate chemical potentials and standard chemical potentials—for example as in eqn (1.30). In reality, as will be explained and emphasised in Chapters 2 and 4, the assumption of ideality is unlikely to be true for the case of electrolyte solutions. It is then necessary to use activities rather than concentrations. An ion, i, has a chemical potential,

$$\mu_i = \mu_i^{\ominus} + RT\ln a_i \tag{1.35}$$

where a_i is the activity of species i, instead of

$$\mu_i = \mu_i^{\ominus} + RT\ln[i] \tag{1.36}$$

for an ideal solution.

It follows that in the previously determined Nernst equations we should more strictly replace the concentration of i, [i] by the activity of i, a_i. The following expressions result:

Fe^{2+}/Fe^{3+} redox couple:

$$\phi_M - \phi_S = \Delta\phi^{\ominus} - \frac{RT}{F}\ln\left\{\frac{a_{Fe^{2+}}}{a_{Fe^{3+}}}\right\} \tag{1.37}$$

H_2/H^+ couple:

$$\phi_M - \phi_S = \Delta\phi^{\ominus} + \frac{RT}{F}\ln\left\{\frac{a_{H^+}}{P_{H_2}^{1/2}}\right\} \tag{1.38}$$

Cl^-/Cl_2 couple:

$$\phi_M - \phi_S = \Delta\phi^\ominus + \frac{RT}{F}\ln\left\{\frac{P_{Cl_2}^{1/2}}{a_{Cl^-}}\right\} \quad (1.39)$$

Ag/AgCl couple:

$$\phi_M - \phi_S = \Delta\phi^\ominus - \frac{RT}{F}\ln\left\{a_{Cl^-}\right\} \quad (1.40)$$

In the rest of this chapter we will use activities. At present simply assume that these adequately approximate to concentration. The next chapter will make clear when this approximation is valid and when it is not.

1.9 A general statement of the Nernst equation for an arbitrary potential determining equilibrium

We are now in a position to generalise the arguments of the previous two sections. Consider any electrode process for which the potential determining equilibrium is

$$\nu_A A + \nu_B B + ... + e^-(\text{metal}) \rightleftharpoons \nu_C C + \nu_D D + ...$$

The terms ν_J (J = A, B, ..., C, D, ...) are known as stoichiometric coefficients. For example in the hydrogen electrode

$$H^+(aq) + e^-(\text{metal}) \rightleftharpoons \tfrac{1}{2}H_2(g)$$

the coefficients are $\nu_{H^+} = 1$ and $\nu_{H_2} = \frac{1}{2}$. Straightforward application of our electrochemical potential arguments as shown in Box 3 leads to the following prediction for the interfacial potential difference

$$\phi_M - \phi_S = \Delta\phi^\ominus + \frac{RT}{F}\ln\left\{\frac{a_A^{\nu_A} a_B^{\nu_B} \cdots}{a_C^{\nu_C} a_D^{\nu_D} \cdots}\right\} \quad (1.41)$$

where we have assumed all the species A, B, ..., C, D, ... to be solution

Box 1.3 The Nernst equation derived

Consider the general reaction

$$\nu_A A + \nu_B B + ... + e^-(\text{metal}) \rightleftharpoons \nu_C C + \nu_D D + ...$$

Equating electrochemical potentials of the reactants and products gives

$$\nu_A\bar{\mu}_A + \nu_B\bar{\mu}_B + ... + \bar{\mu}_{e^-} = \nu_C\bar{\mu}_C + \nu_D\bar{\mu}_D + ...$$

Next the electrochemical potentials are related to chemical potentials of the relevant species:

$$\nu_A(\mu_A + z_A F\phi_S) + \nu_B(\mu_B + z_B F\phi_S) + ... (\mu_{e^-} - F\phi_M) = \nu_C(\mu_C + z_C F\phi_S) + \nu_D(\mu_D + z_D F\phi_S)$$

Conservation of electrical charge in the reaction requires that

$$\nu_A z_A + \nu_B z_B + ... -1 = \nu_C z_C + \nu_D z_D + ...$$

Hence,

$$F(\phi_M - \phi_S) = \nu_A\mu_B + \nu_B\mu_B + ... -\nu_C\mu_C - \nu_D\mu_D + \mu_{e^-}$$

But,

$$\mu_i = \mu_i^{\ominus} + RT\ln a_i$$

where $i = $ A, B, ... C, D, ... and so

$$\phi_M - \phi_S = \Delta\phi^{\ominus} + \frac{RT}{F}(\nu_C \ln a_C + \nu_D \ln a_D + ... - \nu_A \ln a_A - \nu_B \ln a_B - ....)$$

or

$$\phi_M - \phi_S = \Delta\phi^{\ominus} + \frac{RT}{F}\left\{\frac{a_C^{\nu_C} a_D^{\nu_D} ....}{a_A^{\nu_A} a_A^{\nu_B} ....}\right\}$$

which is a general statement of the Nernst equation. The term,

$$\Delta\phi^{\ominus} = (\nu_A \mu_A^{\ominus} + \nu_B \mu_B^{\ominus} + ...) - (\nu_C \mu_C^{\ominus} + \nu_D \mu_D^{\ominus} + ...) + \mu_{e^-}$$

is a constant at a fixed temperature and pressure.

phase molecules. For those which are gaseous it will be appreciated that their activities, a_J, should be replaced by partial pressures, P_J and for those which are pure solids that their activity should be taken as, in effect, unity so that no term relating to the solid appears in the logarithmic term in eqn (1.41). The absence of solid activities can be appreciated by reference to Section 1.7 (Ag/AgCl electrode).

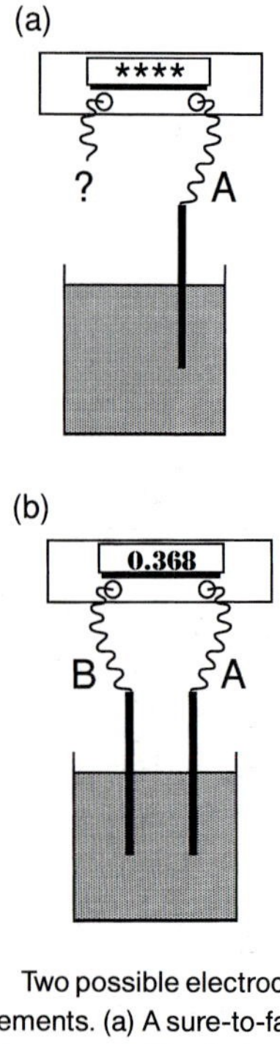

Fig. 1.6 Two possible electrochemical measurements. (a) A sure-to-fail attempt to measure the electrode potential using a single electrode/electrolyte interface. (b) A successful two electrode system employing a reference electrode.

1.10 Measurement of electrode potentials: the need for a reference electrode

We have seen in the preceding sections that the Nernst equation allows us to predict, theoretically, the concentration dependence of the drop in electrical potential, ($\phi_M - \phi_S$), at an electrode solution interface provided we can identify the chemistry of the associated potential-determining equilibrium. It is reasonable at this stage therefore to attempt to compare theory with experiment. However a little thought shows that *it is impossible to measure an absolute value for the potential drop at a single electrode solution interface*. This can be appreciated by considering the experiment shown in Fig. 1.6(a) which displays a sure-to-fail attempt to measure $\phi_M - \phi_S$ using a single electrode/electrolyte ('A') interface with a digital voltameter ('DVM'). The latter has a display which reads out the voltage between the two leads which are shown terminated with connector clips. To record a voltage, the device passes a very tiny current (typically nanoamperes or less) between the connectors to 'probe' the difference in potential. However this clearly will not happen in the case of Fig. 1.6(a) since, whilst a perfectly satisfactory conducting contact can be made with the metal electrode, the link with the solution side of the interface is naturally going to be problematical.

We noted above that for a DVM to measure a voltage a small current must flow through the connector wires between the voltameter terminals. But the only way for current to flow through the experiment shown in

Fig. 1.6(a) is if the connector clip/solution interface, on the left, passes the tiny current of electrons. If no electron passage occurs then we fail to measure the sought quantity. We might introduce the connector clip into the solution but this *generally* does not make electrical contact since *free* electrons will not transfer between the clip and the solution and so the experiment depicted is doomed to failure.

How then are we to pass the necessary current to attempt the sought measurement? If a second electrode, B, is introduced to the solution then the circuit shown in Fig. 1.6(b) may be completed and sensible potentials recorded provided charge is able to pass through the two interfaces—A and B—to permit the DVM to operate. We suppose that the interface, A, under study comprises a system of interest, such as the silver/silver chloride electrode dipping in a solution containing chloride anions:

$$AgCl(s) + e^{-}(metal) \rightleftharpoons Ag(s) + Cl^{-}(aq).$$

In this case electron passage is easy since electrons moving through the silver to the interface can reduce a tiny amount of silver chloride and release chloride ions to carry the current further. Alternatively, if electrons are to leave the silver electrode then a tiny amount of oxidation of the electrode to form silver chloride will provide the charge transfer at the interface: tiny amounts of chloride ions will move towards the silver electrode and electrons will depart from it. If the interface B is simply a connector clip dangling in water then there is no means by which electrons can either be given up to the electrode or accepted from it. The solution near the electrode B must contain chemical species capable of establishing a chemical equilibrium involving the transfer of electrons to or from the electrode. For example, suppose the electrode B is a platinum wire over which hydrogen gas is bubbled and that the solution contains protons. In this case the equilibrium

$$H^{+}(aq) + e^{-}(metal) \rightleftharpoons \tfrac{1}{2}H_2(gas)$$

will be established local to the electrode and the almost infinitesimal interconversion of protons and hydrogen gas can provide a mechanism for the electron transfer through the interface. This situation is depicted in Fig. 1.7 where the solution is shown containing hydrochloric acid. The latter is, of course dissociated into protons and chloride ions. The former assist electron transfer at one interface where the H_2/H^+ couple operates and the latter at the second interface where the Ag/AgCl couple is established. Under these conditions, the DVM is able to pass the nanoampere current necessary to measure the potential between the two electrodes.

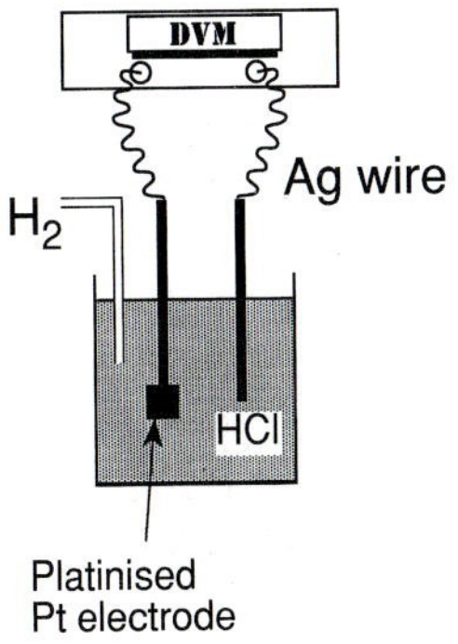

Fig. 1.7 An experimental cell capable of successfully measuring the potential between the two electrodes depicted.

We have seen that interrogation of an electrode–solution interface of interest requires DVM connectors to contact *two* metal electrodes as in Fig. 1.6(b). One of these will make the interface of interest inside the solution; this is labelled A. The second electrode is labelled B. We note that with the arrangement shown in Fig. 1.6(b) the DVM does not measure a quantity $\phi_M - \phi_S$ but rather the difference in potential between the two connectors corresponding to

$$\Delta\phi = (\phi_{\text{Metal A}} - \phi_S) - (\phi_{\text{Metal B}} - \phi_S) = \phi_{\text{Metal A}} - \phi_{\text{Metal B}} \quad (1.42)$$

Recognise two things. First, we have failed to make an absolute measurement of $\phi_{\text{Metal A}} - \phi_S$ and one can now begin to see the impossibility *in general* of such a measurement. Second what we have been able to measure is the *difference* between two terms of the form $(\phi_M - \phi_S)$ as emphasised by eqn (1.42).

Suppose we wish to investigate experimentally a particular electrode system such as one of the three examples covered in Section 1.7; we are forced into the following strategy. For the reasons explained above it is necessary to locate two electrodes in the cell, as depicted in Fig. 1.6(b). Obviously one of these will correspond to the system of interest under 'test'. The purpose of the second electrode is to act as a '*reference electrode*'. The cell is described by the shorthand notation:

Reference Electrode | Solution | Test Electrode

where the vertical line (|) notates a boundary between two separate phases. The measured potential of the cell is given by

$$\text{Measured potential} = (\phi_{\text{test}} - \phi_{\text{solution}}) - (\phi_{\text{reference}} - \phi_{\text{solution}}) \quad (1.43)$$

Suppose the reference electrode operates in such a way that the quantity $(\phi_{\text{reference}} - \phi_{\text{solution}})$ is kept constant. Then

$$\text{Measured potential} = (\phi_{\text{test}} - \phi_{\text{solution}}) + \text{a constant} \quad (1.44)$$

This again tells us that the absolute value of the single electrode quantity $(\phi_{\text{test}} - \phi_{\text{solution}})$ cannot be measured. However if $(\phi_{\text{reference}} - \phi_{\text{solution}})$ is a fixed quantity it follows that any *changes* in $(\phi_{\text{test}} - \phi_{\text{solution}})$ appear directly as changes in the measured potential. Thus if $(\phi_{\text{test}} - \phi_{\text{solution}})$ alters by, say, half a Volt, then the measured potential changes by exactly the same amount. In this way a reference electrode provides us with a method for studying the test electrode but restricts us to knowing about *changes* in the potential of this electrode. However, since this is the best we can possibly achieve, the approach outlined is invariably adopted and when measurements of electrode potentials ('potentiometric measurements') are described throughout the rest of this book, two electrodes—a reference electrode and the electrode of interest—will always be involved in the experiment as shown in Fig. 1.7.

The discussion in this section has introduced the idea of a reference electrode as a device which maintains a fixed value of its potential relative to the solution phase, $(\phi_{\text{reference}} - \phi_{\text{solution}})$, and so facilitates potentiometric measurements of another electrode system *relative* to the reference electrode. This requirement of a fixed value of $(\phi_{\text{reference}} - \phi_{\text{solution}})$ dictates that the reference electrode has certain special properties to ensure that this potential value does indeed stay fixed. In particular any successful reference electrode will display the following properties.

- The chemical composition of the electrode and the solution to which the electrode is directly exposed must be held fixed. This is because the reference electrode potential will be established by some potential-determining equilibrium and the value of this potential will depend on the relative concentrations of the chemical species involved. If these concentrations change the electrode potential also changes. Thus if an AgCl/Ag electrode were used as a reference electrode then

$$(\phi_{\text{reference}} - \phi_{\text{solution}}) = \Delta\phi - \frac{RT}{F}\{\ln a_{\text{Cl}^-}\} \tag{1.45}$$

and it can be appreciated that the chloride ion concentration must be fixed in order for the reference electrode to provide a constant value of $(\phi_{\text{reference}} - \phi_{\text{solution}})$.

- One very important consequence of the requirement for a fixed chemical composition is that it would be disastrously unwise to pass a large electric current through the reference electrode since the passage of this current would induce electrolysis to take place and this would inevitably perturb the concentrations of the species involved in the potential-determining equilibrium. Thus the passage of current through an AgCl/Ag electrode might reduce AgCl to metallic silver so liberating chloride anions if electrons were passed *to* the electrode or, alternatively, the formation of more AgCl at the expense of silver metal and chloride ions if electrons were *removed* from the electrode. In either case the chloride concentration would be changed, and along with it the contribution of the reference electrode to the measured potential.
- It is also experimentally desirable that potential term $(\phi_{\text{reference}} - \phi_{\text{solution}})$ attains its thermodynamic equilibrium value *rapidly*. In other words the potential determining equilibrium should display *fast electrode kinetics*.

The currents involved in measuring a cell potential using a DVM do not cause measurable electrolysis.

1.11 The standard hydrogen electrode

The preceding section has identified the essential characteristics of any reference electrode. Whilst a considerable variety of potentially suitable electrodes are available, for the sake of unambiguity, a single reference electrode has been (arbitrarily) selected for reporting electrode potentials. Thus by convention electrode potentials are quoted for the 'half cell' of interest measured against a standard hydrogen electrode (SHE). This is shown in Fig. 1.8 which shows a complete cell which would measure the Fe^{2+}/Fe^{3+} couple relative to the SHE.

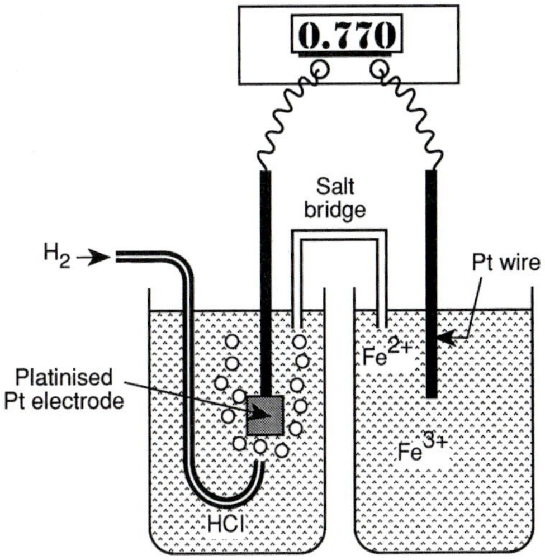

Fig. 1.8 The standard hydrogen electrode employed to measure the potential of the Fe^{2+}/Fe^{3+} couple.

It is interesting to see how the standard hydrogen electrode fulfils the criteria noted in the previous section for a sucessful reference electrode.

- In the *standard* hydrogen electrode the pressure of hydrogen gas is fixed at one atmosphere and the concentration of protons in the aqueous hydrochloric acid is exactly 1.18 M. (We shall see in the next chapter that this proton concentration corresponds precisely to unity activity of H^+.) The temperature is fixed at 298 K (25 °C).
- In Fig. 1.8 the potential between the two electrodes is shown as being measured by means of a digital voltameter (DVM). This device draws essentially negligible current so that no electrolysis of the solution occurs during the measurement and the concentrations of H_2 and H^+ are not perturbed (nor those of Fe^{2+} or Fe^{3+}).
- The reference electrode is fabricated from platinised platinum rather than bright platinum metal to ensure fast electrode kinetics. The purpose of depositing a layer of fine platinum black onto the surface of the platinum is to provide catalytic sites which ensure that the potential determining equilibrium

$$H^+\ (aq) + e^-\ (metal) \rightleftharpoons \tfrac{1}{2}H_2(g)$$

is *rapidly* established. In the absence of this catalysis, on a bright platinum electrode the electrode kinetics are sluggish and cannot be guaranteed to establish the desired electrode potential.

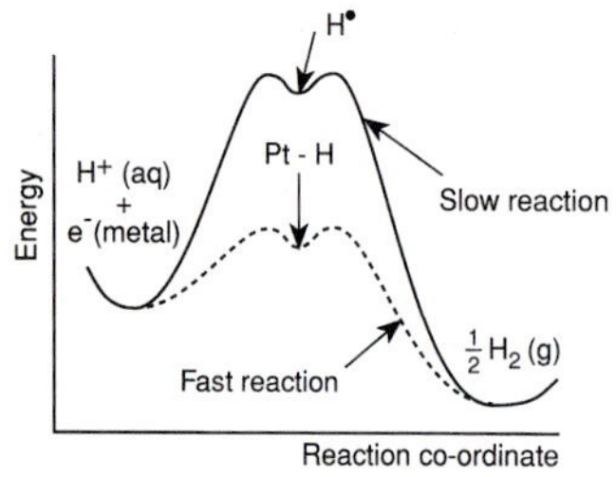

The way platinum black catalyses the H^+/H_2 equilibrium is by providing 'adsorption sites' for the hydrogen atoms, $H^{\cdot}$, formed as intermediates. The adsorption sites permit chemical bonding of the atoms with the electrode surface so stabilising the intermediate and hence speeding up the electrode kinetics.

Consideration of the cell shown in Fig. 1.8 suggests that the potential difference measured by the DVM will correspond to

$$\Delta\phi = (\phi_{\text{Pt wire}} - \phi_{\text{solution}}) - (\phi_{\text{reference}} - \phi_{\text{solution}}) \tag{1.46}$$

Now each of the bracketed terms is given by the Nernst equation for the appropriate electrode. In the case of the platinum wire exposed to the Fe^{2+}/Fe^{3+} solution,

$$(\phi_{\text{Pt wire}} - \phi_{\text{solution}}) = \Delta\phi^{\ominus}_{Fe^{2+}/Fe^{3+}} - \frac{RT}{F}\ln\left\{\frac{a_{Fe^{2+}}}{a_{Fe^{3+}}}\right\} \tag{1.47}$$

For the reference electrode,

$$(\phi_{\text{reference}} - \phi_{\text{solution}}) = \Delta\phi^{\ominus} + \frac{RT}{F}\ln\left\{\frac{a_{H^+}}{P^{1/2}_{H_2}}\right\} \tag{1.48}$$

The measured potential is

$$\Delta\phi = \phi_{\text{Pt wire}} - \phi_{\text{reference}}$$

and can be obtained from eqns (1.47) and (1.48) by simple subtraction:

$$\Delta\phi = (\phi_{\text{wire}} - \phi_{\text{solution}}) - (\phi_{\text{reference}} - \phi_{\text{solution}})$$

so

$$\Delta\phi = \Delta\phi^{\ominus}_{Fe^{2+}/Fe^{3+}} - \Delta\phi^{\ominus}_{H_2/H^+} + \frac{RT}{F}\ln\left\{\frac{a_{Fe^{3+}}P^{1/2}_{H_2}}{a_{Fe^{2+}}a_{H^+}}\right\} \tag{1.49}$$

or

$$E = E^{\ominus} + \frac{RT}{F}\ln\left\{\frac{a_{Fe^{3+}}P^{1/2}_{H_2}}{a_{Fe^{2+}}a_{H^+}}\right\} \tag{1.50}$$

where

$$E^\ominus = \left(\Delta\phi^\ominus_{Fe^{2+}/Fe^{3+}} - \Delta\phi^\ominus_{H_2/H^+}\right) \quad (1.51)$$

The value of $E^\ominus$ is known as the 'standard electrode potential' of the Fe^{2+}/Fe^{3+} couple. This is the measured potential of the cell shown in Fig. 1.8 when the hydrogen electrode is standard ($a_{H^+} = 1$; $P_{H_2} = 1$ atm) and when all the chemical species contributing to the potential determining equilibrium are present at unity activity (or, approximately speaking, concentration) so that

$$a_{Fe^{2+}} = a_{Fe^{3+}} = 1.$$

It is worth emphasising again that it is implicit in the above that negligible current is drawn in the measurement of the potential since this would lead to a change of composition in the cell. Finally inspection of Fig. 1.8 shows that the two half cells—the Fe^{2+}/Fe^{3+} and the H_2/H^+ couples—are physically separated by means of a salt bridge. This is a tube containing an aqueous solution of potassium chloride which places the two half cells in electrical contact. One purpose of the salt bridge is to stop the two different solutions required for the two half cells from mixing. Otherwise, for example, the platinum electrode forming part of the standard hydrogen reference electrode would be exposed to the Fe^{2+}/Fe^{3+} potential-determining equilibrium and its potential accordingly disrupted.

We will discuss salt bridges in Chapter 4. In particular it will be seen that the chemical composition of the solution inside the bridge is crucial. Aqueous potassium chloride is acceptable but potassium sulphate is not!

1.12 Standard electrode potentials

In general the Standard electrode potential (SEP) of any system ('couple' or 'half cell') is defined as the measured potential difference between the two electrodes of a cell in which the potential of the electrode of interest is measured relative to the SHE and in which *all* the chemical species contributing to the potential determining equilibria at each electrode are present at a concentration corresponding to unit activity in the case of a solution phase species, or to unit pressure in the case of a gas.

As a further example the consider the SEP of the Cu/Cu^{2+} couple. This quantity is the measured potential between the two electrodes shown in Fig. 1.9. The activity (approximately concentration) of the copper (II)

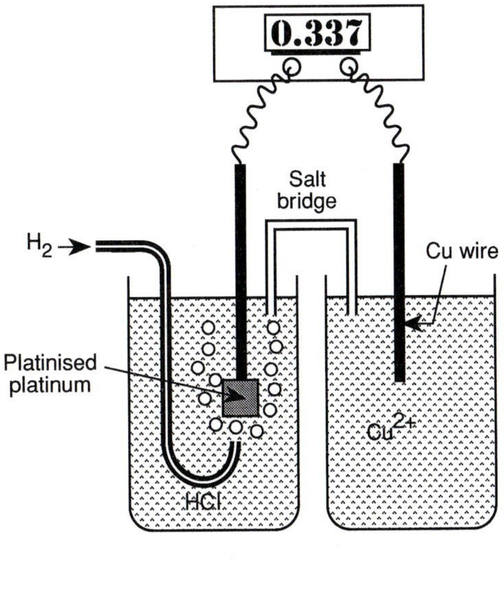

Fig. 1.9 The standard hydrogen electrode employed to measure the standard electrode potential of the Cu/Cu^{2+} couple. Note that $P_{H2} = 1$ atm, $a_{H^+} = 1$ and $a_{Cu^{2+}} = 1$.

ions is unity and a salt bridge is used again to prevent the solutions of the two half cells from mixing.

Hitherto when we have described an electrochemical cell we have drawn a picture such as is shown in Figs 1.8 or 1.9. However this is a tedious business which can be avoided by the introduction of a shorthand notation which avoids the need for a pictorial representation. For example the cell shown in Fig. 1.9 is described by the following notation:

$$\mathrm{Pt} \mid \mathrm{H_2(g)}\ (P = 1\ \mathrm{atm}) \mid \mathrm{H^+(aq)}\ (a = 1) \parallel \mathrm{Cu^{2+}(aq)}\ (a = 1) \mid \mathrm{Cu}$$

where the symbol $\parallel$ denotes the salt bridge. The single vertical lines, $|$, denote boundaries between two separate phases.

The SEP of the Cu/Cu^{2+} couple is given by,

$$E^{\ominus}_{\mathrm{Cu/Cu^{2+}}} = \phi_{\mathrm{Cu}} - \phi_{\mathrm{Pt}} \tag{1.52}$$

When measured the potential difference between the copper and platinum electrodes is found to be 0.34 V with the copper electrode positively charged and the platinum electrode negatively charged. In writing down potentials a convention is essential so that the correct polarity is assigned to the cell. This is done as follows: *with reference to a cell diagram, the potential is that of the right hand electrode relative to that of the left hand electrode, as the diagram is written down.* Thus for

$$\mathrm{Cu} \mid \mathrm{Cu^{2+}(aq)}\ (a = 1) \parallel \mathrm{H^+(aq)}\ (a = 1) \mid \mathrm{H_2(g)}\ (P = 1\ \mathrm{atm}) \mid \mathrm{Pt}$$
$$E_{\mathrm{cell}} = -0.34\ V = \phi_{\mathrm{Pt}} - \phi_{\mathrm{Cu}}$$

However for

$$\mathrm{Pt} \mid \mathrm{H_2\ (g)}\ (P = 1\ \mathrm{atm}) \mid \mathrm{H^+(aq)}\ (a = 1),\ \mathrm{Cu^{2+}(aq)}\ (a = 1) \mid \mathrm{Cu}$$
$$E_{\mathrm{cell}} = +0.34\ V = \phi_{\mathrm{Cu}} - \phi_{\mathrm{Pt}}$$

Tabulations of standard electrode potentials for all are available for very many half cells. Table 1 gives a limited number of examples. In Table 1 the potentials are measured relative to a standard hydrogen electrode at 25 °C. For example the third entry refers to the cell

$$\mathrm{Pt} \mid \mathrm{H_2(g)}\ (P = 1\ \mathrm{atm}) \mid \mathrm{H^+(aq)}\ (a = 1),\ \mathrm{Br^-(aq)}\ (a = 1) \mid \mathrm{AgBr} \mid \mathrm{Ag}$$
$$E^{\ominus}_{\mathrm{Ag/AgBr}} = +0.07 = \phi_{\mathrm{Ag}} - \phi_{\mathrm{Pt}}$$

Such tables are valuable since they allow us to predict the potential of *any cell* formed from any *pair of half cells.* For example the cell,

$$\mathrm{Cu} \mid \mathrm{Cu^{2+}(aq)}\ (a = 1) \parallel \mathrm{Zn^{2+}\ (aq)}\ (a = 1) \mid \mathrm{Zn}$$

as, shown in Fig. 1.10, has a potential of

$$E^{\ominus}_{\mathrm{cell}} = \phi_{\mathrm{Zn}} - \phi_{\mathrm{Cu}} \tag{1.53}$$

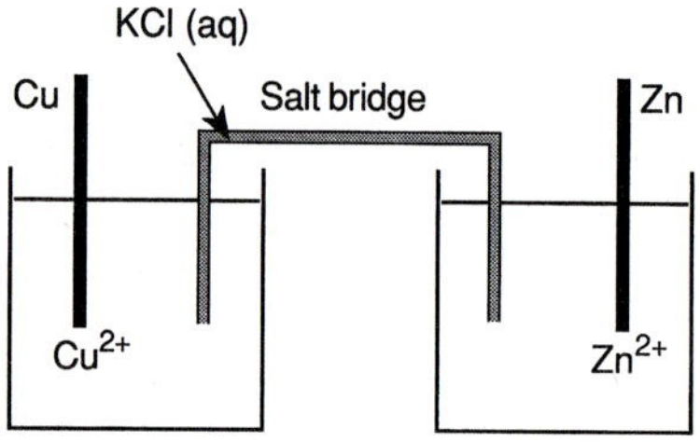

Fig. 1.10 An electrochemical experiment measuring the potential between the Cu^{2+}/Cu and Zn^{2+}/Zn half cells. Note that $a_{\mathrm{Cu^{2+}}} = a_{\mathrm{Zn^{2+}}} = 1$.

Table 1.1 Standard Electrode Potentials (25°C)

	$E^{\ominus}$/V		$E^{\ominus}$/V
$Ag^+ + e^- \rightarrow Ag$	+0.80	$\frac{1}{2}Hg_2SO_4 + e^- \rightarrow Hg + \frac{1}{2}SO_4^{2-}$	+0.62
$Ag^{2+} + e^- \rightarrow Ag^+$	+1.98	$\frac{1}{2}I_2 + e^- \rightarrow I^-$	+0.54
$AgBr + e^- \rightarrow Ag + Br^-$	+0.07	$\frac{1}{2}I_3^- + e^- \rightarrow \frac{3}{2}I^-$	+0.53
$AgCl + e^- \rightarrow Ag + Cl^-$	+0.22	$In^+ + e^- \rightarrow In$	–0.13
$AgI + e^- \rightarrow Ag + I^-$	–0.15	$\frac{1}{2}In^{3+} + e^- \rightarrow \frac{1}{2}In^+$	–0.44
$\frac{1}{3}Al^{3+} + e^- \rightarrow \frac{1}{3}Al$	–1.68	$\frac{1}{3}In^{3+} + e^- \rightarrow \frac{1}{3}In$	–0.34
$Au^+ + e^- \rightarrow Au$	+1.83	$K^+ + e^- \rightarrow K$	–2.93
$\frac{1}{3}Au^{3+} + e^- \rightarrow \frac{1}{3}Au$	+1.52	$Li^+ + e^- \rightarrow Li$	–3.04
$\frac{1}{2}Ba^{2+} + e^- \rightarrow \frac{1}{2}Ba$	–2.92	$\frac{1}{2}Mg^{2+} + e^- \rightarrow \frac{1}{2}Mg$	–2.36
$\frac{1}{2}Be^{2+} + e^- \rightarrow \frac{1}{2}Be$	–1.97	$\frac{1}{2}Mn^{2+} + e^- \rightarrow \frac{1}{2}Mn$	–1.18
$\frac{1}{2}Br_2(l) + e^- \rightarrow Br^-$	+1.06	$Mn^{3+} + e^- \rightarrow Mn^{2+}$	+1.51
$\frac{1}{2}Br_2(aq) + e^- \rightarrow Br^-$	+1.09	$\frac{1}{2}MnO_2 + 2H^+ + e^- \rightarrow \frac{1}{2}Mn^{2+} + H_2O$	+1.23
$\frac{1}{2}BrO^- + \frac{1}{2}H_2O + e^- \rightarrow \frac{1}{2}Br^- + OH^-$	+0.76	$MnO_4^- + e^- \rightarrow MnO_4^{2-}$	+0.56
$HOBr + H^+ + e^- \rightarrow \frac{1}{2}Br_2 + H_2O$	+1.60	$NO_3^- + 2H^+ + e^- \rightarrow NO_2 + H_2O$	+0.80
$\frac{1}{2}BrO_4^- + H^+ + e^- \rightarrow \frac{1}{2}BrO_3^- + \frac{1}{2}H_2O$	+1.85	$\frac{1}{3}NO_3^- + \frac{4}{3}H^+ + e^- \rightarrow \frac{1}{3}NO + \frac{2}{3}H_2O$	+0.96
$\frac{1}{2}Ca^{2+} + e^- \rightarrow \frac{1}{2}Ca$	–2.84	$\frac{1}{2}NO_3^- + \frac{1}{2}H_2O + e^- \rightarrow \frac{1}{2}NO_2^- + OH^-$	+0.01
$\frac{1}{2}Cd(OH)_2 + e^- \rightarrow \frac{1}{2}Cd + OH^-$	–0.82	$Na^+ + e^- \rightarrow Na$	–2.71
$\frac{1}{2}Cd^{2+} + e^- \rightarrow \frac{1}{2}Cd$	–0.40	$\frac{1}{2}Ni^{2+} + e^- \rightarrow \frac{1}{2}Ni$	–0.26
$\frac{1}{3}Ce^{3+} + e^- \rightarrow \frac{1}{3}Ce$	–2.34	$\frac{1}{2}Ni(OH)_2 + e^- \rightarrow \frac{1}{2}Ni + OH^-$	–0.72
$Ce^{4+} + e^- \rightarrow Ce^{3+}$	+1.72	$\frac{1}{2}NiO_2 + 2H^+ + e^- \rightarrow \frac{1}{2}Ni^{2+} + H_2O$	+1.59
$\frac{1}{2}Cl_2 + e^- \rightarrow Cl^-$	+1.36	$\frac{1}{4}O_2 + \frac{1}{2}H_2O + e^- \rightarrow OH^-$	+0.40
$\frac{1}{2}ClO^- + \frac{1}{2}H_2O + e^- \rightarrow \frac{1}{2}Cl^- + OH^-$	+0.89	$\frac{1}{4}O_2 + H^+ + e^- \rightarrow \frac{1}{2}H_2O$	+1.23
$HOCl + H^+ + e^- \rightarrow \frac{1}{2}Cl_2 + H_2O$	+1.63	$O_2 + e^- \rightarrow O_2^-$	–0.33
$ClO_3^- + 2H^+ + e^- \rightarrow ClO_2 + H_2O$	+1.17	$\frac{1}{2}O_2 + \frac{1}{2}H_2O + e^- \rightarrow \frac{1}{2}HO_2^- + \frac{1}{2}OH^-$	–0.08
$\frac{1}{2}ClO_4^- + H^+ + e^- \rightarrow \frac{1}{2}ClO_3^- + \frac{1}{2}H_2O$	+1.20	$O_2 + H^+ + e^- \rightarrow HO_2$	–0.13
$\frac{1}{2}Co^{2+} + e^- \rightarrow \frac{1}{2}Co$	–0.28	$\frac{1}{2}O_2 + H^+ + e^- \rightarrow \frac{1}{2}H_2O_2$	+0.70
$Co^{3+} + e^- \rightarrow Co^{2+}$	+1.92	$\frac{1}{2}Pb^{2+} + e^- \rightarrow \frac{1}{2}Pb$	–0.13
$Co(NH_3)_6^{3+} + e^- \rightarrow Co(NH_3)_6^{2+}$	+0.06	$\frac{1}{2}PbO_2 + 2H^+ + e^- \rightarrow \frac{1}{2}Pb^{2+} + H_2O$	+1.70
$\frac{1}{2}Cr^{2+} + e^- \rightarrow \frac{1}{2}Cr$	–0.90	$\frac{1}{2}PbSO_4 + e^- \rightarrow \frac{1}{2}Pb + \frac{1}{2}SO_4^{2-}$	–0.36
$\frac{1}{3}Cr^{3+} + e^- \rightarrow \frac{1}{3}Cr$	–0.74	$\frac{1}{2}Pt^{2+} + e^- \rightarrow \frac{1}{2}Pt$	+1.19
$Cs^+ + e^- \rightarrow Cs$	–2.92	$Rb^+ + e^- \rightarrow Rb$	–2.93
$Cu^+ + e^- \rightarrow Cu$	+0.52	$\frac{1}{2}S + e^- \rightarrow \frac{1}{2}S^{2-}$	–0.48
$\frac{1}{2}Cu^{2+} + e^- \rightarrow \frac{1}{2}Cu$	+0.34	$\frac{1}{2}SO_2(aq) + \frac{1}{2}H^+ + e^- \rightarrow \frac{1}{4}S_2O_3^{2-} + \frac{1}{4}H_2O$	–0.40
$Cu^{2+} + e^- \rightarrow Cu^+$	+0.16	$\frac{1}{4}SO_2(aq) + H^+ + e^- \rightarrow \frac{1}{4}S + \frac{1}{2}H_2O$	+0.50
$CuCl + e^- \rightarrow Cu + Cl^-$	+0.12	$\frac{1}{2}S_4O_6^{2-} + e^- \rightarrow S_2O_3^{2-}$	+0.08
$\frac{1}{2}Cu(NH_3)_4^{2+} + e^- \rightarrow \frac{1}{2}Cu + 2NH_3$	–0.00	$\frac{1}{2}SO_4^{2-} + \frac{1}{2}H_2O + e^- \rightarrow \frac{1}{2}SO_3^{2-} + OH^-$	–0.94
$\frac{1}{2}F_2 + e^- \rightarrow F^-$	+2.87	$SO_4^{2-} + 2H^+ + e^- \rightarrow \frac{1}{2}S_2O_6^{2-} + H_2O$	–0.25
$\frac{1}{2}Fe^{2+} + e^- \rightarrow \frac{1}{2}Fe$	–0.44	$\frac{1}{2}S_2O_8^{2-} + e^- \rightarrow SO_4^{2-}$	+1.96
$\frac{1}{3}Fe^{3+} + e^- \rightarrow \frac{1}{3}Fe$	–0.04	$\frac{1}{2}Sn^{2+} + e^- \rightarrow \frac{1}{2}Sn$	–0.14
$Fe^{3+} + e^- \rightarrow Fe^{2+}$	+0.77	$\frac{1}{2}Sn^{4+} + e^- \rightarrow \frac{1}{2}Sn^{2+}$	+0.15
$Fe(CN)_6^{3-} + e^- \rightarrow Fe(CN)_6^{4-}$	+0.36	$\frac{1}{2}Sr^{2+} + e^- \rightarrow \frac{1}{2}Sr$	–2.89
$\frac{1}{2}Fe(CN)_6^{4-} + e^- \rightarrow \frac{1}{2}Fe + 3CN^-$	–1.16	$\frac{1}{2}Ti^{2+} + e^- \rightarrow \frac{1}{2}Ti$	–1.63
$H^+ + e^- \rightarrow \frac{1}{2}H_2$	0.00	$Ti^{3+} + e^- \rightarrow Ti^{2+}$	–1.37
$H_2O + e^- \rightarrow \frac{1}{2}H_2 + OH^-$	–0.83	$TiO^{2+} + 2H^+ + e^- \rightarrow Ti^{3+} + H_2O$	+0.10
$H_2O_2 + H^+ + e^- \rightarrow OH + H_2O$	+0.71	$Tl^+ + e^- \rightarrow Tl$	–0.34
$\frac{1}{2}H_2O_2 + H^+ + e^- \rightarrow H_2O$	+1.76	$\frac{1}{2}V^{2+} + e^- \rightarrow \frac{1}{2}V$	–1.13
$\frac{1}{2}Hg^{2+} + e^- \rightarrow \frac{1}{2}Hg$	+0.80	$V^{3+} + e^- \rightarrow V^{2+}$	–0.26
$\frac{1}{2}Hg_2Cl_2 + e^- \rightarrow Hg + Cl^-$	+0.27	$\frac{1}{2}VO^{2+} + H^+ + e^- \rightarrow \frac{1}{2}V^{2+} + \frac{1}{2}H_2O$	+0.34
$\frac{1}{2}Hg^{2+} + e^- \rightarrow \frac{1}{2}Hg$	+0.86	$VO_2^+ + 2H^+ + e^- \rightarrow VO^{2+} + H_2O$	+1.00
$Hg^{2+} + e^- \rightarrow \frac{1}{2}Hg_2^{2+}$	+0.91	$\frac{1}{2}Zn^{2+} + e^- \rightarrow \frac{1}{2}Zn$	–0.76

which is equivalent to

$$E^{\ominus}_{cell} = E^{\ominus}_{Zn/Zn^{2+}} - E^{\ominus}_{Cu/Cu^{2+}} \tag{1.54}$$

The values of $E^{\ominus}_{Zn/Zn^{2+}}$ and $E^{\ominus}_{Cu/Cu^{2+}}$ can be found from the table and hence the cell potential estimated. The cell potential can also be calculated when the species are of a known activity other than unity. For example the cell

$$Cu \mid Cu^{2+}(aq)\ (a_{Cu^{2+}}) \parallel Zn^{2+}\ (aq)\ (a_{Zn^{2+}}) \mid Zn$$

will have a cell potential given by

$$E_{cell} = \phi_{Zn} - \phi_{Cu} = (\phi_{Zn} - \phi_{solution}) - (\phi_{Cu} - \phi_{solution}) \tag{1.55}$$

which, when the Nernst equation is written for each term in round brackets, gives

$$E_{cell} = \left(\Delta\phi^{\ominus}_{Zn/Zn^{2+}} + \frac{RT}{F}\ln a^{1/2}_{Zn^{2+}}\right) - \left(\Delta\phi^{\ominus}_{Cu/Cu^{2+}} + \frac{RT}{F}\ln a^{1/2}_{Cu^{2+}}\right) \tag{1.56}$$

and this reduces to

$$E_{cell} = \left(\Delta\phi^{\ominus}_{Zn/Zn^{2+}} - \Delta\phi^{\ominus}_{Cu/Cu^{2+}}\right) + \frac{RT}{F}\ln\left\{\frac{a^{1/2}_{Zn^{2+}}}{a^{1/2}_{Cu^{2+}}}\right\} \tag{1.57}$$

If it is noted that

$$E^{\ominus}_{Zn/Zn^{2+}} - E^{\ominus}_{Cu/Cu^{2+}} = (\Delta\phi^{\ominus}_{Zn/Zn^{2+}} - \Delta\phi^{\ominus}_{H^+/H_2}) - (\Delta\phi^{\ominus}_{Cu^{2+}/Cu} - \Delta\phi^{\ominus}_{H^+/H_2}) \tag{1.58}$$

Extensive tabulations of electrode potentials can be found in *Standard Potentials in Aqueous Solution* edited by A. J. Bard, R. Parsons and J. Jordan published by Marcel Dekker, New York; 1985. It contains over 800 pages of densely packed data, testifying to the great value of electrode potentials in summarising solution equilibria and thermodynamic data.

then the equation for the potential difference across the cell transforms to

$$E_{cell} = E^{\ominus}_{Zn/Zn^{2+}} - E^{\ominus}_{Cu/Cu^{2+}} + \frac{RT}{2F}\ln\left\{\frac{a_{Zn^{2+}}}{a_{Cu^{2+}}}\right\} \tag{1.59}$$

which is the Nernst equation for the cell. Using values of $E^{\ominus}_{Zn/Zn^{2+}}$ and $E^{\ominus}_{Cu/Cu^{2+}}$ from Table 1.1 permits us now to evaluate the cell potential for any values of the activities, $a_{Zn^{2+}}$ and $a_{Cu^{2+}}$.

1.13 The Nernst equation applied to a general cell

The Nernst equation for *any* electrochemical cell can be obtained simply by following the procedure outlined below which is a generalisation of the strategy followed in the preceding section:-

(i) Write down the cell in the shorthand notation given in Section 1.12.

(ii) Write the reactions at the two electrodes as reductions involving *one* electron only (as in Table 1.1). Writing the reaction as a reduction means that the electron appears in the left hand side of the equation in each case.

(iii) Subtract the reaction at the left hand electrode (in the cell as written down) from the reaction at the right hand electrode to find a 'formal cell reaction':

$$\sum_i \nu_i R_i \rightarrow \sum_j \nu_j P_j \tag{1.60}$$

where R_i represent the reactant species, P_j, the product species and v_i and v_j are their stoichiometric coefficients.

(iv) The resulting Nernst equation is given by

$$E_{cell} = E^{\ominus}_{cell} + \frac{RT}{F} \ln \frac{\prod a_{R_i}^{v_i}}{\prod a_{P_j}^{v_j}} \quad (1.61)$$

If any of the reactants, R_i or products, P_j, are pure solids, their activities are effectively unity — see Section 1.9.

where

$$E^{\ominus}_{cell} = E^{\ominus}_{right} - E^{\ominus}_{left}$$

and $E^{\ominus}_{right}$ and $E^{\ominus}_{left}$ are the standard electrode potentials of the half cell reactions for the right and left hand electrodes as drawn.

Next we illustrate the above procedure for the cell

$$Cd \mid Cd^{2+}(aq)\ (a_{Cd^{2+}}) \parallel Pb^{2+}\ (aq)\ (a_{Pb^{2+}}) \mid Pb$$

Step (ii) gives the reaction at the right hand electrode to be:

$$\tfrac{1}{2}Pb^{2+}(aq) + e^{-}(metal) \rightarrow \tfrac{1}{2}Pb(s)$$

At the left hand electrode the reaction is:

$$\tfrac{1}{2}Cd^{2+}(aq) + e(metal)^{-} \rightarrow \tfrac{1}{2}Cd(s)$$

Following step (iii) and subtracting gives:

$$\tfrac{1}{2}Pb^{2+}(aq) + \tfrac{1}{2}Cd(s) \rightarrow \tfrac{1}{2}Pb(s) + \tfrac{1}{2}Cd^{2+}(aq)$$

which is the 'formal cell reaction'. Step (iv) leads to:

$$E_{cell} = E^{\ominus}_{cell} + \frac{RT}{F} \ln \left\{ \frac{a_{Pb^{2+}}^{1/2}}{a_{Cd^{2+}}^{1/2}} \right\} \quad (1.62)$$

where

$$E^{\ominus}_{cell} = E^{\ominus}_{Pb/Pb^{2+}} - E^{\ominus}_{Cd/Cd^{2+}}$$

Examination of Table 1.1 shows that $E^{\ominus}_{Pb/Pb^{2+}} = -0.126$ V and $E^{\ominus}_{Cd/Cd^{2+}} = --0.403 Vsothat$

$$E^{\ominus}{}_{cell} = (-0.126) - (-0.403) = +0.277 \text{ V}$$

Therefore the Nernst equation for the cell is:

$$E_{cell} = 0.277 + \frac{RT}{2F} \ln \left\{ \frac{a_{Pb^{2+}}}{a_{Cd^{2+}}} \right\} \quad (1.63)$$

This equation predicts the cell potential for *any* values of the activities $a_{Pb^{2+}}$ and $a_{Cd^{2+}}$.

It should be noted that the *formal* cell reaction, as introduced in step (iii), depends upon how the cell is written down in step (i). For example, for the cell

$$Cd \mid Cd^{2+}(aq)\ (a_{Cd^{2+}}) \parallel Pb^{2+}\ (aq)\ (a_{Pb^{2+}}) \mid Pb$$

the value of $E^{\ominus}_{cell}$ is $+0.277$ V and the formal cell reaction is

$$\tfrac{1}{2}Pb^{2+}(aq) + \tfrac{1}{2}Cd(s) \rightarrow \tfrac{1}{2}Pb(s) + \tfrac{1}{2}Cd^{2+}(aq)$$

In contrast, for the cell

$$\mathrm{Pb} \mid \mathrm{Pb}^{2+}(\mathrm{aq})\ (a_{\mathrm{Pb}^{2+}}) \parallel \mathrm{Cd}^{2+}\ (\mathrm{aq})\ (a_{\mathrm{Cd}^{2+}}) \mid \mathrm{Cd}$$

the value of $E^{\ominus}_{\mathrm{cell}}$ is –0.277 V and the formal cell reaction is:

$$\tfrac{1}{2}\mathrm{Pb(s)} + \tfrac{1}{2}\mathrm{Cd}^{2+}(\mathrm{aq}) \rightarrow \tfrac{1}{2}\mathrm{Pb}^{2+}(\mathrm{aq}) + \tfrac{1}{2}\mathrm{Cd(s)}$$

It is thus helpful to distinguish the *formal* cell reaction from the *spontaneous* cell reaction. The latter is the reaction that would occur if the cell were shortcircuited. That is, the two electrodes were directly connected to each other, for example using a conducting wire. The nature of the spontaneous cell reaction can be readily deduced since, in reality, electrons will flow from a negative electrode to a positive electrode through an external circuit as is illustrated in the scheme below:

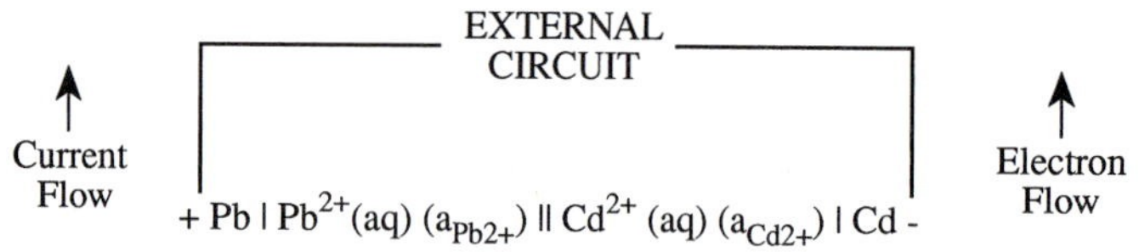

Notice the procedure given above predicts the cadmium electrode to be negatively charged and the lead electrode to carry a positive charge. Electrons therefore pass from the cadmium electrode through the external circuit to the lead electrode. This implies that oxidation occurs at the right hand electrode:

$$\tfrac{1}{2}\mathrm{Cd} \rightarrow \tfrac{1}{2}\mathrm{Cd}^{2+} + e^-$$

and reduction occurs at the left hand electrode:

$$\tfrac{1}{2}\mathrm{Pb}^{2+} + e^- \rightarrow \tfrac{1}{2}\mathrm{Pb}$$

It follows that the spontaneous cell reaction is:

$$\tfrac{1}{2}\mathrm{Pb}^{2+} + \tfrac{1}{2}\mathrm{Cd} \rightarrow \tfrac{1}{2}\mathrm{Pb} + \tfrac{1}{2}\mathrm{Cd}^{2+}$$

This therefore is the formal reaction for the cell

$$\mathrm{Cd} \mid \mathrm{Cd}^{2+}(\mathrm{aq})\ (a_{\mathrm{Cd}^{2+}}) \parallel \mathrm{Pb}^{2+}\ (\mathrm{aq})\ (a_{\mathrm{Pb}^{2+}}) \mid \mathrm{Pb}$$

but is opposite of that we deduced earlier for the cell

$$\mathrm{Pb} \mid \mathrm{Pb}^{2+}(\mathrm{aq})\ (a_{\mathrm{Pb}^{2+}}) \parallel \mathrm{Cd}^{2+}\ (\mathrm{aq})\ (a_{\mathrm{Cd}^{2+}}) \parallel \mathrm{Cd}$$

In general the spontaneous cell reaction that occurs when the two electrodes of any cell are shortcircuited may be established using the protocol set out at the beginning of the present section, and tables of electrode potentials, to decide which electrode is positively charged and which negatively charged. Electron flow in the external current will always be from the negative to the positive electrode so that an oxidation process will occur at the former and a reduction at the latter.

1.14 The relation of electrode potentials to the thermodynamics of the cell reaction

Before studying this section the reader might wish to read Boxes 1.4 to 1.6 to refresh their recollection of some aspects of chemical thermodynamics.

Box 1.4 Entropy and the second law

The entropy of a system, S, is a measure of its disorder. Changes in entropy can arise through the addition of heat to a system. If an infinitesimal amout of heat, dQ, is added *reversibly* then the resulting increase in entropy is

$$dS = dQ/T$$

where T is the temperature of the system. The entropy increases ($dS > 0$) since the added heat promotes disorder in the system—the molecules will move with a slightly larger range of velocities and possess a slightly broader range of energies.

The second law of thermodynamics tells us that the entropy of the universe is steadily increasing. Expressed another way changes can only occur if, as a result, the universal entropy increases.

We have noted in previous sections that cell potentials are measured under conditions where only negligible current is drawn. Consequently electrons are transferred from one half cell to the other under essentially *thermodynamically reversible conditions.* With this in mind let us consider the cell illustrated in the scheme below.

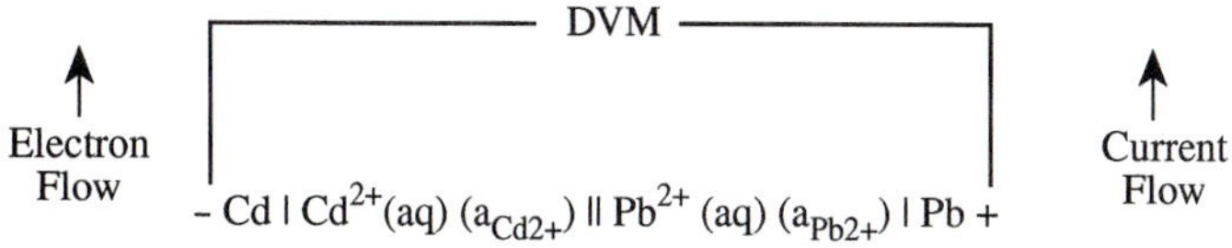

If an amount, dn moles, of electrons flow from the negative electrode to the positive electrode, then dn moles of the following reaction will occur.

$$\tfrac{1}{2}Pb^{2+}(aq) + \tfrac{1}{2}Cd(s) \rightarrow \tfrac{1}{2}Pb(s) + \tfrac{1}{2}Cd^{2+}(aq)$$

Associated with this, will be a change dG in the Gibbs free energy of the cell. As explained in Boxes 1.4 to 1.6:

$$dG = dw_{\text{additional}} \qquad (1.63)$$

where $dw_{\text{additional}}$ corresponds to the work done (other than 'PdV work') in the process. In the above scheme the only contribution to this quantity is the work done in transferring the charge ($-Fdn$ coulombs) through the external circuit across a potential difference of E_{cell} volts. It follows that

$$dG = dw_{\text{additional}} = (-Fdn)E_{\text{cell}} \qquad (1.64)$$

Therefore for each mole of electrons transferred

$$\Delta G = -FE_{\text{cell}} \qquad (1.65)$$

where ΔG refers to the reaction

$$\tfrac{1}{2}Pb^{2+}(aq) + \tfrac{1}{2}Cd(s) \rightarrow \tfrac{1}{2}Pb(s) + \tfrac{1}{2}Cd^{2+}(aq)$$

Enthalpy changes are heat changes at constant pressure so that $dQ_{system} = dH_{system}$

Box 1.5 Gibbs free energy

The second law of thermodynamics tells us that a reaction can only occur if there is an *increase* in the *entropy of the universe* ($S_{universe}$). Mathematically this can be stated as:

$$dS_{universe} > 0$$

If we consider a system and its surroundings,

$$dS_{system} + dS_{surroundings} > 0 \qquad \text{(i)}$$

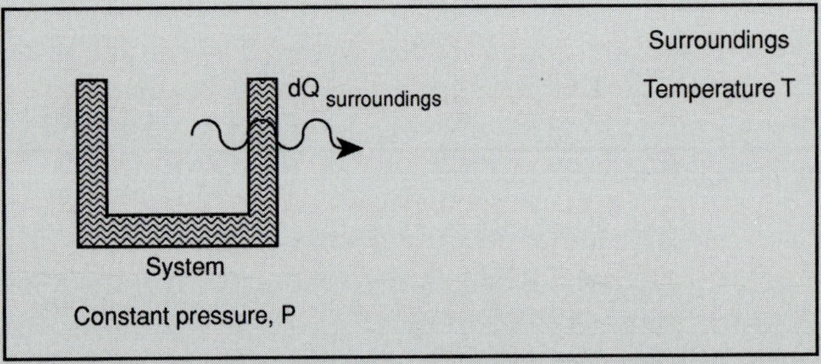

The surroundings, at a temperature T are generally large enough to exchange heat with the system *reversibly*, and so from the definition of entropy

$$dS_{surroundings} = dQ_{surroundings}/T$$

where $dQ_{surroundings}$ is the heat gained by the surroundings as a result of the reaction in the system. But the heat gained by the surroundings is equal to that lost by the system, so

$$dQ_{surroundings} = -dQ_{system} = -dH_{system}$$

where H_{system} is the enthalpy of the system. Therefore

$$dS_{surroundings} = -dH_{system}/T$$

Substituting this into inequality (i) we obtain

$$TdS_{system} - dH_{system} > 0$$

or

$$dG_{system} < 0 \text{ (at constant temperature)} \qquad \text{(ii)}$$

where

$$G = H - TS$$

G is the famous Gibbs free energy. It follows that (i) and (ii) are equivalent forms of the second law and *an increase in the universal entropy corresponds to a decrease in the free energy of the system.* Therefore changes in a system can only occur if, as a result, the free energy of the system decreases.

If the cell components are at unit activity then,

$$\Delta G^{\ominus} = -FE^{\ominus} \tag{1.66}$$

It can therefore be seen that the measurement of cell potentials provides information about free energy changes. Furthermore, since

$$dG = VdP - SdT \tag{1.67}$$

it can be concluded that

$$\left(\frac{\partial G}{\partial T}\right)_P = -S \tag{1.68a}$$

and

$$\left(\frac{\partial \Delta G}{\partial T}\right)_P = -\Delta S \tag{1.68b}$$

So

$$F\left(\frac{\partial E}{\partial T}\right)_P = \Delta S \quad \text{and} \quad F\left(\frac{\partial E^{\ominus}}{\partial T}\right)_P = \Delta S^{\ominus} \tag{1.69}$$

Combining the above and

$$\Delta H^{\ominus} = \Delta G^{\ominus} + T\Delta S^{\ominus} \tag{1.70}$$

Box 1.6 Some further thermodynamics

The enthalpy H is given by

$$H = U + PV$$

where U is the internal energy, P the pressure and V the volume. By expanding the term for enthalpy the Gibbs free energy can be expressed as

$$G = H - TS = U + PV - TS$$

Therefore at constant pressure and temperature

$$dG = dU + PdV - TdS \tag{i}$$

Changes, dU, in the internal energy arise from heat changes, dQ, and from the work done on or by the system, dw. Under *reversible* conditions

$$\begin{aligned} dU &= dQ_{rev} + dw_{rev} \\ &= TdS + dw_{rev} \qquad (ii) \\ &= TdS + (-PdV + dw_{additional}) \end{aligned}$$

where $(-PdV)$ is the work done by the system in expanding against the atmosphere and dw_{added} is any other sort of work. By combining (i) and (ii) we obtain

$$dG = dw_{additional}$$

under reversible conditions. The quantity $dw_{additional}$ corresponds to the work done on the system, other than 'PdV work'. This might be the work done in transferring charge through an electrical circuit between two different electrical potentials.

gives

$$\Delta H^{\ominus} = -FE^{\ominus} + TF\left(\frac{\partial E^{\ominus}}{\partial T}\right)_P \tag{1.71}$$

From this it can be seen that the entropy and enthalpy of a cell reaction can be obtained from the cell potential and its variation with temperature.

1.15 Standard electrode potentials and the direction of chemical reactions

Knowledge of the standard electrode potential of a cell allows us to establish, thermodynamically, the direction of the corresponding cell reaction. Suppose that analysis of the cell using the protocol in Section 1.13 leads to the following formal cell reaction,

$$\sum_i \nu_i R_i \rightarrow \sum_j \nu_j P_j \tag{1.72}$$

The standard potential for the cell is given by

$$E^{\ominus}_{\mathrm{cell}} = E^{\ominus}_{\mathrm{right}} - E^{\ominus}_{\mathrm{left}} \tag{1.73}$$

Now if we note,

$$\Delta G^{\ominus} = -FE^{\ominus} = -RT \ln K \tag{1.74}$$

we obtain

$$E^{\ominus} = (RT/F) \ln K \tag{1.75}$$

Box 1.7 The relationship between $\Delta G^{\ominus}$ and K

Consider a general chemical reaction

$$\nu_A A + \nu_B B + \ldots \rightleftharpoons \nu_C C + \nu_D D + \ldots$$

At equilibrium the sum of the chemical potentials of the reactants will equal that of the products:

$$\nu_A\mu_A + \nu_B\mu_B + \ldots = \nu_C\mu_C + \nu_D\mu_D + \ldots$$

But

$$\mu_i = \mu_i^{\ominus} + RT\ln a_i$$

or

$$\mu_i = \mu_i^{\circ} + RT\ln P_i$$

depending on whether i is a solution or gas phase species: a_i is the activity of species i and P_i is the partial pressure. It follows that, in solution

$$\{\nu_C\mu_C^{\ominus} + \nu_D\mu_D^{\ominus} + \ldots - \nu_A\mu_A^{\ominus} - \nu_B\mu_B^{\ominus} - \ldots\}$$

$$= -\nu_C RT\ln a_C - \nu_D RT\ln a_D - \ldots + \nu_A RT\ln a_A + \nu_B RT\ln a_B + \ldots$$

$$= -RT\,(\ln a_C{}^{\nu_C} + \ln a_D{}^{\nu_D} + \ldots - \ln a_A{}^{\nu_A} - \ln a_B{}^{\nu_B} - \ldots$$

$$= -RT\ln\left\{\frac{a_C^{\nu_C}.a_D^{\nu_D}\ldots}{a_A^{\nu_A}.a_B^{\nu_B}\ldots}\right\}$$

This equation may be re-written

$$\Delta G^{\ominus} = -RT\ln K_a \qquad \text{(i)}$$

where

$$K_a = \frac{a_C^{\nu_C} . a_D^{\nu_D} \ldots}{a_A^{\nu_A} . a_B^{\nu_B} \ldots}$$

and

$$\Delta G^{\ominus} = \{\nu_C\mu_C^{\ominus} + \nu_D\mu_D^{\ominus} + \ldots - \nu_A\mu_A^{\ominus} - \nu_B\mu_B^{\ominus} - \ldots\}$$

For a gas phase process an analogous argument leads to

$$\Delta G^{\ominus} = -RT\ln K_p \qquad \text{(ii)}$$

where

$$K_p = \frac{P_C^{\nu_C} . P_D^{\nu_D} \ldots}{P_A^{\nu_A} . P_B^{\nu_B} \ldots}$$

Equations (i) and (ii) show how the equilibrium constant for a reaction is related to the free energy change, $\Delta G^{\ominus}$, for the reaction. If $\Delta G^{\ominus}$ is very exothermic ($\Delta G^{\ominus} < 0$) then the equilibrium constant favours the products ($K_p > 0$ or $K_a > 0$). Alternatively if it is endothermic then it favours the reactants ($\Delta G^{\ominus} > 0$; $K_p < 0$ or $K_a < 0$). The following table emphasises this point.

$\Delta G^{\ominus}$ changes and the corresponding *K* values

$\Delta G^{\ominus}$	K	*Comment*
-50 kJ mol^{-1}	6×10^{8}	*Products Favoured*
-10 kJ mol^{-1}	57	
0	1	
$+10$ kJ mol^{-1}	0.02	
$+50$ kJ mol^{-1}	2×10^{-9}	*Reactants Favoured*

Notice that

$$\Delta G^{\ominus} = \{\nu_C\mu_C^{\ominus} + \nu_D\mu_D^{\ominus} + \ldots - \nu_A\mu_A^{\ominus} - \nu_B\mu_B^{\ominus} - \ldots\}$$

$$= \left\{\begin{matrix}\text{Gibbs Free Energy} \\ \text{of Pure Products C, D, ...,} \\ \text{in their standard states}\end{matrix}\right\} - \left\{\begin{matrix}\text{Gibbs Free Energy} \\ \text{of Pure Products A, B, ...,} \\ \text{in their standard states}\end{matrix}\right\}$$

For example,

(*a*) The reaction of dinitrogen tetroxide to form nitrogen (IV) oxide

$$N_2O_4(g) \rightleftharpoons 2NO_2(g)$$

$$K_p = \frac{P_{NO_2}^2}{P_{N_2O_4}}$$

$$\Delta G^{\ominus} = 2\mu^{\ominus}{}_{NO_2(g)} - \mu^{\ominus}{}_{N_2O_4(g)}$$

= {Gibbs free energy of 2 moles of pure $NO_2(g)$ at one atmosphere pressure and specified temperature} **minus** {Gibbs free energy of 1 mole of pure $N_2O_4(g)$ at one atmosphere pressure and specified temperature}

The specified temperature will be that selected to define the standard state. Often this is 298 K.

(*b*) The bromination of phenol to form tribromophenol

$$C_6H_5OH(aq) + 3Br_2(aq) \rightarrow 3HBr\ (aq) +$$ 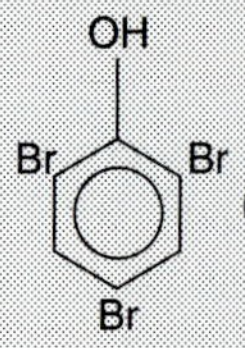

(aq)

TBP

$$K = \frac{a_{TBP} . a_{HBr}^3}{a_{C_6H_5OH} . a_{Br_2}^3}$$

$$\Delta G^{\ominus} = 3\mu^{\ominus}_{HBr(aq)} + \mu^{\ominus}_{TBP(aq)} - 3\mu^{\ominus}_{Br_2(aq)} - \mu^{\ominus}_{C_6H_5OH(aq)}$$

= {Gibbs Free Energy of 3 Moles of HBr
in a solution in which it has unit activity
plus that of 1 mole of TBP
in a solution in which it has unit activity;
both at a specified temperature}
minus {Gibbs Free Energy of 3 Moles of Br_2
in a solution in which it has unit activity
plus that of 1 mole of phenol
in a solution in which it has unit activity;
both at a specified temperature}

Again the temperature defining the standard state needs to be specified, for example $\Delta G^{\ominus}_{298}$.

and so we can conclude that if $E^{\ominus}$ is greater than zero K will be greater than one, and if $E^{\ominus}$ is negative K will be less than unity for the cell reaction. For example if we consider the SEP of a metal/metal ion couple as noted in the cell below

$$\text{Pt} \mid H_2(g)\ (P = 1\ \text{atm}) \mid H^+(aq)\ (a = 1),\ M^{n+}(aq)\ (a = 1) \mid M \quad (1.76)$$

the formal cell reaction will be

$$\tfrac{1}{n}M^{n+}(aq) + \tfrac{1}{2}H_2(g) \rightarrow \tfrac{1}{n}M(s) + H^+(aq) \quad (1.77)$$

and so the standard electrochemical potential of the metal/metal ion couple indicates whether or not the metal will react with H^+(aq) to give hydrogen gas. Thus for example in the case of gold, we consider the following cell

$$\text{Pt} \mid H_2(g)\ (P = 1\ \text{atm}) \mid H^+(aq)\ (a = 1)\ ,\ Au^+(aq)\ (a = 1) \mid Au$$

The standard electrode potential is +1.83 V and so for the reaction

$$Au^+(aq) + \tfrac{1}{2}H_2(g) \rightarrow Au(s) + H^+(aq)$$

the standard reaction free energy is

$$\Delta G^{\ominus} = -1.83F$$

It follows that gold will *not* react with acid (H^+) under standard conditions ($a_{H^+} = 1$) to form hydrogen gas. Conversely, considering the reaction

$$Li^+(aq) + \tfrac{1}{2}H_2(g) \rightarrow Li(s) + H^+(aq)$$

the standard electrode potential for the Li/Li^+ couple is –3.04 V so that for the above reaction

$$\Delta G^{\ominus} = +3.04F$$

showing that the reaction of Li with acid is strongly favourable in thermodynamic terms. The inertness of gold and the reactivity of lithium in aqueous acid predicted in this way will be in line with the readers chemical experience.

Generalising the above, it can be seen that if a metallic element M has a standard electrode potential for the M/M^{n+} couple which is negative, then it is possible for M to react with acid under standard conditions ($a_{H^+} = 1 \approx [H^+]/\text{mol dm}^{-3}$) to evolve hydrogen. If however the standard electrode potential is positive then the reaction is impossible thermodynamically when exposed to aqueous acid. We expect, on examining Table 1.1, that Li, Na and Zn will all evolve H_2 but Au, Pt and Cu will not.

As a further example we note that copper is capable of forming both mono-valent and di-valent ions, Cu^+ and Cu^{2+}, in aqueous solution. If we consider the disproportionation reaction of copper(I),

$$2Cu^+(aq) \rightleftharpoons Cu(s) + Cu^{2+}(aq)$$

This can be broken down into two separate cell reactions.

$$Cu^+(aq) + e^-(metal) \rightarrow Cu(s)$$

and

$$Cu^{2+}(aq) + e^-(metal) \rightarrow Cu^+(aq)$$

Turning to Table 1 we see that $E^{\ominus}$ for the former reaction is +0.52 V and that for the latter is +0.16 V. It follows that for the reaction

$$Cu^+(aq) + \tfrac{1}{2}H_2(g) \rightleftharpoons Cu(s) + H^+(aq)$$

$$\Delta G^{\ominus} = -0.52F \text{ kJ mol}^{-1}$$

Likewise

$$Cu^{2+}(aq) + \tfrac{1}{2}H_2(g) \rightleftharpoons Cu^+(aq) + H^+(aq)$$

$$\Delta G^{\ominus} = -0.16F \text{ kJ mol}^{-1}$$

The two reactions may be subtracted to give the disproportionation reaction:

$$\begin{array}{ll} & Cu^+(aq) + \tfrac{1}{2}H_2(g) \rightleftharpoons Cu(s) + H^+(aq) \\ \text{minus} & Cu^{2+}(aq) + \tfrac{1}{2}H_2(g) \rightleftharpoons Cu^+(aq) + H^+(aq) \\ \hline \text{gives} & 2Cu^+(aq) \rightleftharpoons Cu(s) + Cu^{2+}(aq) \end{array}$$

for which,

$$\Delta G^{\ominus} = (-0.52F) - (-0.16F) = -0.36F$$

and so

$$K = \frac{a_{Cu^+}\, a_{Cu^+}}{a_{Cu^{2+}}} = 8.3 \times 10^{-7} \text{mol dm}^{-3}$$

at the standard temperature of 298 K. We conclude that the disproportionation reaction is very likely to occur. Indeed copper(I) disproportionates very rapidly in water, with a lifetime typically of less than one second, forming metallic copper and copper(II) ions.

Box 1.8 Mixed equilibrium constants

In Box 1.7 we considered equilibria in the gas phase and in solution. Often, however, reactions involve species in more than one phase. For example

$$AgCl\ (s) \rightleftharpoons Ag^{+}\ (aq) + Cl^{-}\ (aq)$$

$$CaCO_3\ (s) \rightleftharpoons CaO\ (s) + CO_2\ (g)$$

$$Fe^{2+}(aq) + \tfrac{1}{2}Cl_2(g) \rightleftharpoons Fe^{3+}(aq) + Cl^{-}\ (aq).$$

Let us examine equilibria in a system in which there are different phases and focus on a general reaction:

$$aA(g) + \ldots\ bB(aq) + \ldots cC(s) \rightleftharpoons dD(g) + \ldots\ eE(aq) + \ldots\ fF(s) + \ldots$$

At equilibrium the sum of the reactant and product chemical potentials will be equal so

$$a\mu_{A(g)} + \ldots\ b\mu_{B(aq)} + \ldots\ c\mu_{C(s)} + \ldots \rightleftharpoons d\mu_{D(g)} + \ldots\ e\mu_{E(aq)} + \ldots f\mu_{F(s)} + \ldots$$

But,

$$\mu_{A(g)} = \mu_A^{o} + RT\ln P_A$$

$$\mu_{B(aq)} = \mu_B^{\ominus} + RT\ln[B(aq)]$$

$$\mu_{D(g)} = \mu_D^{o} + RT\ln P_D$$

and

$$\mu_{E(aq)} = \mu_E^{\ominus} + RT\ln[E(aq)]$$

However

$$\mu_{C(s)} = \mu_C^{\square} \text{ and } \mu_{F(s)} = \mu_F^{\square}$$

where $\mu_i^{\square}$ is the Gibbs free energy of one mole of pure, solid i under standard conditions (say one atmosphere pressure and 298 K). Note the different superscripts o and $\ominus$ for gases and solution phase species respectively.

It follows that

$$\Delta G^{\ominus} = \{d\mu_D^{\circ} + ... e\mu_E^{\ominus} + ... f\mu_F^{\square} + ...\} - \{a\mu_A^{\circ} + ... b\mu_B^{\ominus} + ... c\mu_C^{\square} + ...\}$$

$$= -dRT\ln P_D - .... eRT\ln[\mathrm{E(aq)}] - .... + aRT\ln P_A + ... bRT\ln[\mathrm{B(aq)}] + ...$$

$$= -RT\{\ln P_D^d + ... \ln[\mathrm{E}]^e + ... -\ln P_A^a - .... -\ln[\mathrm{B}]^b - ....\}$$

$$= -RT\ln\left\{\frac{P_D^d......[\mathrm{E}]^e......}{P_A^a......[\mathrm{B}]^b......}\right\} = -RT\ln K$$

where

$$K = \frac{P_D^d......[\mathrm{E}]^e......}{P_A^a......[\mathrm{B}]^b......}$$

The equilibrium constant contains reference to the gaseous species A,...., D, ... and the species B, ..., E ...., dissolved in solution but does *not* contain the solid species C, ..., F, ... Perusal of the above argument shows that the absence results from the fact that the chemical potentials μ_C and μ_F are fixed at a specified temperature. In contrast for A and D, and B and E the partial pressures and the concentrations (activities) influence the chemical potentials.

Returning to the equilibria first considered, the appropiate equilibium constants can now be seen to be

$$K = [\mathrm{Ag}^+]\,[\mathrm{Cl}^-] \text{ (or } K = a_{\mathrm{Ag}^+} \,.\, a_{\mathrm{Cl}^-}),$$

$$K = P_{\mathrm{CO}_2}$$

and

$$K = \frac{[\mathrm{Cl}^-][\mathrm{Fe}^{3+}]}{[\mathrm{Fe}^{2+}]P_{\mathrm{Cl}_2}^{\frac{1}{2}}}$$

Finally we consider the case where the solvent—for example water—is also a reactant. Consider the auto-ionisation of water,

$$2\mathrm{H_2O\ (l)} \rightleftharpoons \mathrm{H_3O^+\ (aq)} + \mathrm{OH^-\ (aq)}$$

Now, except in very concentrated solutions, to a very good approximation

$$\mu_{\mathrm{H_2O\ (l)}} \approx \mu_{\mathrm{H_2O\ (l)}}^{\triangledown}$$

it follows that the appropriate equilibrium constant is

$$K_W = [\mathrm{H_3O^+}]\,[\mathrm{OH^-}] = 10^{-14}\ \mathrm{mol^2\ dm^{-6}}\ (\text{at } 25^{\circ}\mathrm{C})$$

Notice that eqn (1.8a) rather than (1.8b) is , as here, usually applied to the solvent. Likewise for the dissociation of weak acids, such as ethanoic acid,

$$\mathrm{CH_3COOH\ (aq)} + \mathrm{H_2O\ (l)} \rightleftharpoons \mathrm{CH_3COO^-\ (aq)} + \mathrm{H_3O^+\ (aq)}$$

the acid dissociation constant is

$$K = \frac{[H_3O^+][CH_3COO^-]}{[CH_3COOH]}$$

or

$$K = \frac{a_{H_3O^+} a_{CH_3COO^-}}{a_{CH_3COOH}}$$

1.16 Standard electrode potentials and disproportionation

Section 1.15 showed how standard electrode potentials could be used to predict the disproportionation of copper(I) in aqueous solution. The conclusions can be generalised to the reaction

$$(a+b)M^{x+}(aq) \rightleftharpoons aM^{(x+b)+}(aq) + bM^{(x-a)+}(aq) \qquad (1.78)$$

If we consider the separate reactions

$$\frac{1}{b}M^{(x+b)+}(aq) + \frac{1}{2}H_2(g) \rightleftharpoons \frac{1}{b}M^{x+}(aq) + H^+(aq) \qquad (1.79)$$

and

$$\frac{1}{a}M^{x+}(aq) + \frac{1}{2}H_2(g) \rightleftharpoons \frac{1}{a}M^{(x-a)+}(aq) + H^+(aq) \qquad (1.80)$$

for which

$$\Delta G^\ominus = -FE^\ominus{}_{M^{x+}/M^{(x+b)+}} \qquad (1.81)$$

and

$$\Delta G^\ominus = -FE^\ominus{}_{M^{(x-a)+}/M^{x+}} \qquad (1.82)$$

respectively. Thus

(*ab*) times $\quad \frac{1}{a}M^{x+}(aq) + \frac{1}{2}H_2(g) \rightleftharpoons \frac{1}{a}M^{(x-a)+}(aq) + H^+(aq)$

minus (*ab*) times $\quad \frac{1}{b}M^{x+b}(aq) + \frac{1}{2}H_2(g) \rightleftharpoons \frac{1}{b}M^{x+}(aq) + H^+(aq)$

gives $\quad (a+b)M^{x+}(aq) \rightleftharpoons aM^{(x+b)+}(aq) + bM^{(x-a)+}(aq) \qquad (1.83)$

so that

$$\Delta G^\ominus = -abF\,\{E^\ominus{}_{M^{(x-a)+}/M^{x+}} - E^\ominus{}_{M^{x+}/M^{(x+b)+}}\} \qquad (1.84)$$

It follows that the $\Delta G^\ominus$ will be negative, and the disproportionation favourable, if

$$E^\ominus{}_{M^{x+}/M^{(x+b)+}} < E^\ominus{}_{M^{(x-a)+}/M^{x+}}$$

In the case of copper(I)

$$E^\ominus{}_{Cu^+/Cu^{2+}} < E^\ominus{}_{Cu/Cu^+}$$

and thus disproportionation is favourable. In contrast

$$E^{\ominus}{}_{Fe^{2+}/Fe^{3+}} < E^{\ominus}{}_{Fe/Fe^{2+}}$$

so the reaction

$$3Fe^{2+}(aq) \rightleftharpoons 2Fe^{3+}(aq) + Fe(s)$$

is not favoured: the addition of iron to a solution of iron(III) leads to the formation of iron(II).

1.17 Standard electrode potentials and pH

Consider the disproportionation of bromine

$$3Br_2(aq) + 3H_2O(l) \rightarrow BrO^-{}_3(aq) + 6H^+(aq) + 5Br^-(aq)$$

$pH = -\log_{10} a_{H_3O^+}$

Since

$$E^{\ominus}_{Br_2/BrO_3^-} = +1.48\ V$$

and

$$E^{\ominus}{}_{Br^-/Br_2} = +1.06\ V$$

it follows from eqn 1.84 in the previous section that for the reaction written above

$$\Delta G^{\ominus} = -F.5.\{1.06\text{–}1.48\}$$
$$= +0.42 \times 5F$$
$$= +2.10\ F\ \text{kJ mol}^{-1}$$

Hence,

$$K = \frac{a^6_{H^+} \cdot a_{BrO_3^-} \cdot a^5_{Br^-}}{a^3_{Br_2}} = \exp\left(\frac{-2.10F}{RT}\right)$$
$$= 3.06 \times 10^{-36}$$

Thus at pH = 0, where $a_{H^+} = 1$ the disproportionation is unfavourable. However at pH = 9, we can deduce that

$$\frac{a_{BrO_3^-} \cdot a^5_{Br^-}}{a^3_{Br_2}} = 3.06 \times 10^{-36} \times (10^9)^6$$
$$= 3.06 \times 10^{18}$$

so that in weakly basic solution the disproportionation becomes thermodynamically possible. Whenever protons or hydroxide ions appear for particular redox couples in Table 1.1, the equilibria involved in these couples will be sensitive to the solution pH and by varying this quantity the equilibrium may shifted in favour of products, or reactants.

1.18 Thermodynamics versus kinetics

The previous sections have given illustration of the use of electrode potentials in predicting the position of chemical equilibria. The predictions are, however, subject to *kinetic* limitations. That is, if a reaction is thermodynamically feasible, does the reaction 'go' at a reasonable rate? Consider the reaction

$$\tfrac{1}{2}\mathrm{Mg(s)} + \mathrm{H_2O(l)} \rightarrow \tfrac{1}{2}\mathrm{H_2(g)} + \mathrm{OH^-(aq)} + \tfrac{1}{2}\mathrm{Mg^{2+}(aq)}$$

Table 1.1 tells us that at 298 K

$$E^{\ominus}_{\mathrm{Mg/Mg^{2+}}} = -2.36 \text{ V}$$

This implies that for the reaction

$$\tfrac{1}{2}\mathrm{Mg^{2+}(aq)} + \tfrac{1}{2}\mathrm{H_2(g)} \rightarrow \tfrac{1}{2}\mathrm{Mg(s)} + \mathrm{H^+(aq)}$$

the standard free energy change at 298 K is

$$\Delta G^{\ominus} = +2.36F \text{ kJ mol}^{-1}$$

$E^{\ominus}_{\mathrm{H_2,OH^-/H_2O}}$ implies the cell,
Pt | H_2(g) (P = 1 atm) | H^+(aq) (a = 1) || OH^-(aq) (a = 1)| H_2 (g) (P = 1 atm) | Pt
so that the potential-determining equilibria are:
Right Hand Electrode:
$e^-(\mathrm{metal}) + \mathrm{H_2O(l)} \rightarrow \tfrac{1}{2}\mathrm{H_2(g)} + \mathrm{OH^-(aq)}$
Left Hand Electrode:
$e^-(\mathrm{metal}) + \mathrm{H^+(aq)} \rightarrow \tfrac{1}{2}\mathrm{H_2(g)}$
giving the formal cell reaction implied in the text.

Likewise, since

$$E^{\ominus}_{\mathrm{H_2,\,OH^-/H_2O}} = -0.83 \text{ V}$$

for the reaction

$$\mathrm{H_2O(l)} \rightarrow \mathrm{OH^-(aq)} + \mathrm{H^+(aq)}$$

the corresponding energy change is

$$\Delta G^{\ominus} = +0.83F \text{ kJ mol}^{-1}$$

It follows that for the reaction

$$\tfrac{1}{2}\mathrm{Mg(s)} + \mathrm{H_2O(l)} \rightarrow \tfrac{1}{2}\mathrm{H_2(g)} + \mathrm{OH^-(aq)} + \tfrac{1}{2}\mathrm{Mg^{2+}(aq)}$$

$$\Delta G^{\ominus} = (+0.83\text{–}2.36)\ F \text{ kJ mol}^{-1} = -1.53F \text{ kJ mol}^{-1}$$

so that when magnesium metal is dipped into water the evolution of hydrogen gas is thermodynamically viable. However, in practice, little or no reaction is observed since a thin film of magnesium oxide, present on the metal surface, prevents the reaction taking place. The oxide layer *passivates* the metal. Similar reasons explain the lack of reaction of aluminium or titanium with water.

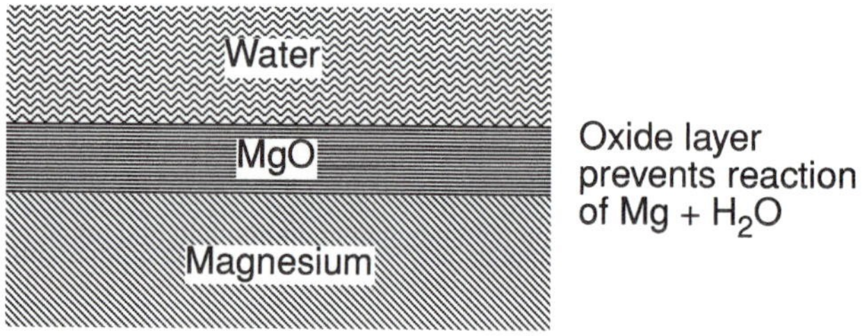

Fig. 1.11 The passivation of metallic magnesium.

To generalise, the use of standard electrode potentials can tell us if a reaction is thermodynamically viable. However if a process is predicted to be possible, standard electrode potentials tell us nothing about the likely *rate of reaction.*

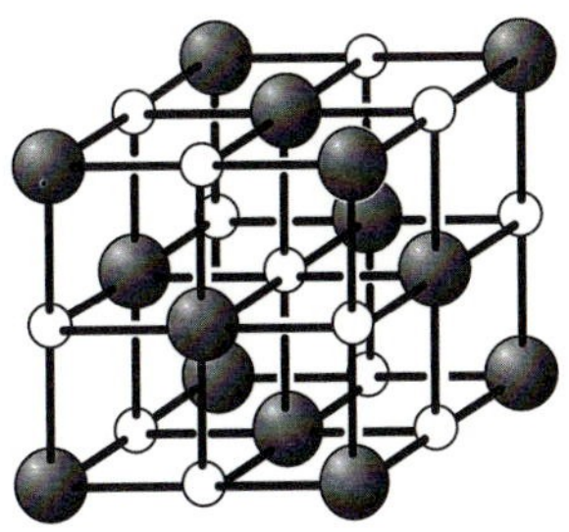

Magnesium oxide adopts the sodium chloride structure.

1.19 The measurement of standard electrode potentials

This chapter has introduced the physical origin of electrode potentials and shown the great value of standard electrode potentials in deducing thermodynamic quantities such as free energy, entropy and enthalpy changes for reactions in solution together with their equilibrium constants. Specifically tabulations of data such as Table 1.1 enable the deduction of the point of equilibrium attained in *any* reaction which can be stated as the sum of any two 'half reactions' from the table. It will be appreciated that the accurate and valuable measurements of standard electrode potentials is a topic of large importance in solution chemistry. It is to this topic that we next turn. However before tackling the experimental measurement of standard electrode potentials we need to briefly consider two additional topics. The first concerns the relationship between concentration and activities and the second addresses how different ions move about in solution. These issues are examined in the next two chapters before the theme of measuring standard electrode potentials is resumed in Chapter 4.

Bibliographical note: Walther Hermann Nernst (1864–1941).
W. H. Nernst, the son of a provincial judge, was born on 25 June 1864 in Briesen, West Prussia. He began as a physicist, finishing his undergraduate studies in 1886 at Graz where he met Boltzmann.

Nernst's transformation to physical chemist was a result of his time as an assistant to Ostwald in Liepzig, where the group also included Arrhenius and Van't Hoff. It was during this period that Nernst started the work which would finally give rise to the Nernst equation. In 1894 he became the first Professor of Physical Chemistry at Gottingen. By 1900 Nernst had realised that interfacial potential differences between different phases were not individually measurable and thus concluded that electrochemical potentials could only be measured relative to another and proposed the hydrogen electrode as the standard. This insight allowed Nernst to formulate his equation for the potential of a general cell.

In 1905 he became Professor of Physical Chemistry at Berlin. By this time his work had moved away from solution chemistry. In 1920 he received a Nobel prize for his heat theorem work of 1906, which was the first statement of the Third Law of Thermodynamics. After 2 years as the director of the Physikalisch Technisches Reichensalt, Nernst became the Professor of Experimental Physics at Berlin in 1924 and he stayed there until he retired in 1933. He died in November 1941 after spending much time in his last years dedicated to cosmological speculations.

Nernst was undoubtedly a great practical chemist although his friend Einstein noted that his 'theoretical equipment was somewhat elementary, but he mastered it with rare ingenuity'. He also suggested that Nernst's textbook *Physical Chemistry* (a standard from it's publication in 1893 until the 1920s) offered the student 'an abundance of stimulating ideas' but was 'theoretically elementary'. To quote Einstein once again he also had a 'childlike vanity and self-complacency' not to mention a 'egocentric weakness', however he was 'neither a nationalist nor a militarist' judging 'people almost exclusively by their direct success'.

An interesting invention of Nernst that attracted much attention, but did not become widely adopted, was an electric piano in which the player regulated a source of electrical energy which excited the strings to vibration.

2 Allowing for non-ideality: activity coefficients

In this chapter the relationship between *activity* and *concentration* is developed

2.1 Introduction

The chemical potential of an *ideal* solution is given by:

$$\mu_A = \mu_A^{\ominus} + RT \ln[A] \tag{2.1}$$

or

$$\mu_A = \mu_A^{\nabla} + RT \ln x_A \tag{2.2}$$

depending upon whether we choose to deal with concentrations or mole fractions. However for *non-ideal solutions* the chemical potential is given by:

$$\mu_A = \mu_A^{\ominus} + RT \ln a_A \tag{2.3}$$

where a_A is the 'effective' concentration of A in the solution or the *activity of A*. It is related to the concentration of the solution by the activity coefficient γ:

$$a_A = \gamma_A [A] \tag{2.4}$$

As defined, the activity has units of moles dm^{-3}. Occasionally, and more rigorously, activities are defined in terms of molalities, m, where $a_A = \gamma_A m_A$. The quantity m_A has units of moles kg^{-1} and, unlike a concentration it is temperature independent. For aqueous solutions $m_A \approx [A]$ since one litre of water has a mass of approximately 1 kilogram. Both definitions will be used in this Primer.

Clearly if γ_A is unity then the solution is ideal. Otherwise the solution is non-ideal and the extent to which γ_A deviates from unity is a measure of the solution's non-ideality. In any solution we usually know [A] but not either a_A or γ_A. However we shall see in this chapter that for the special case of dilute electrolytic solutions it is possible to calculate γ_A. This calculation involves the *Debye–Hückel theory* to which we turn in Section 2.4. It provides a method by which activities may be quantified through a knowledge of the concentration combined with the Debye–Huckel calculation of γ_A. First, however, we consider some relevant results pertaining to ideal solutions and, second in Section 2.3, a general interpretation of γ_A.

2.2 The entropy of mixing: ideal solutions

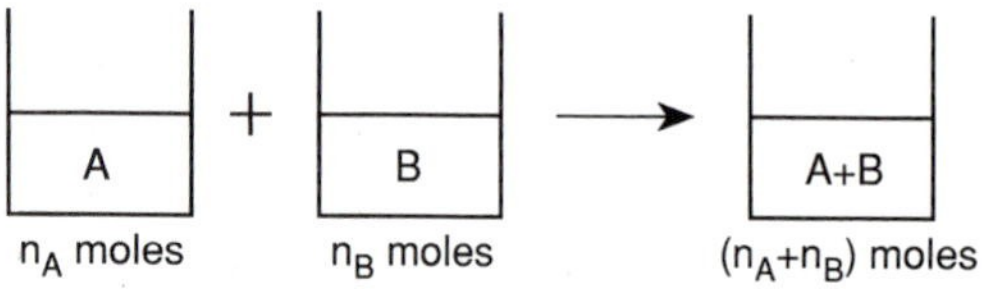

Fig. 2.1 The mixing of two solutions.

Consider what happens when two separate solutions containing n_A and n_B moles of A and B respectively are mixed together. The initial and final free energies will be given by:

$$\text{Initial free energy} = n_A \mu_A^{\nabla} + n_B \mu_B^{\nabla} \quad (2.5)$$

$$\text{Final free energy} = n_A \mu_A + n_B \mu_B \quad (2.6)$$

$$= n_A\mu_A^{\nabla} + n_A RT \ln x_A + n_B \mu_B^{\nabla} + n_B RT \ln x_B \quad (2.7)$$

So, clearly,

$$\text{Change in free energy} = RT\,(n_A \ln x_A + n_B \ln x_B) \quad (2.8)$$

and thus

$$\text{Free energy of mixing per mole of solution} = \frac{RT}{n_A + n_B}(n_A \ln x_A + n_B \ln x_B) \quad (2.9)$$

and so

$$\Delta G_{mix} = RT\,(x_A \ln x_A + x_B \ln x_B) \quad (2.10)$$

where ΔG_{mix} is measured in kJ mol^{-1} and will be negative. The variation of ΔG_{mix} with x_A is sketched in Fig. 2.2.; note that $x_B = 1 - x_A$. The free energy of mixing attains its largest negative value when $x_A = x_B = 0.5$. Now, recognising that

$$dG = VdP - SdT \quad (2.11)$$

we obtain

$$\left(\frac{\partial G}{\partial T}\right)_P = -S \quad \text{or} \quad \left(\frac{\partial \Delta G}{\partial T}\right)_P = -\Delta S \quad (2.12)$$

and so the entropy of mixing is, on differentiating eqn 2.9 with respect to temperature, given by

$$\Delta S_{mix} = -R\,(x_A \ln x_A + x_B \ln x_B) \quad (2.13)$$

The enthalpy of mixing may then be computed

$$\Delta H_{mix} = \Delta G_{mix} + T\Delta S_{mix} = 0. \quad (2.14)$$

We have therefore shown that the enthalpy of mixing is zero for ideal solutions. Physically this can be interpreted in the terms that the intermolecular forces between the molecules in the mixture are equal to those in the pure liquids. In other words A–A, A–B and B–B interactions are the same. Equality of interaction forces is therefore an alternative way of describing an ideal solution

It follows from the above that non-ideality must arise from grossly dissimilar inter-particular forces. A little thought shows we would expect electrolyte solutions to be seriously non-ideal and this is found to be case except for very dilute solutions ($< 10^{-3}$ M). Consider for example an aqueous solution of NaCl. The species in the solution are Na^+ cations, Cl^- anions and dipolar water molecules. Evidently the Na^+– Na^+, Na^+– Cl^-, Cl^-– H_2O, Cl^-– Cl^- and Na^+– H_2O interactions are all very different and we would expect the solution to be non ideal. So in considering

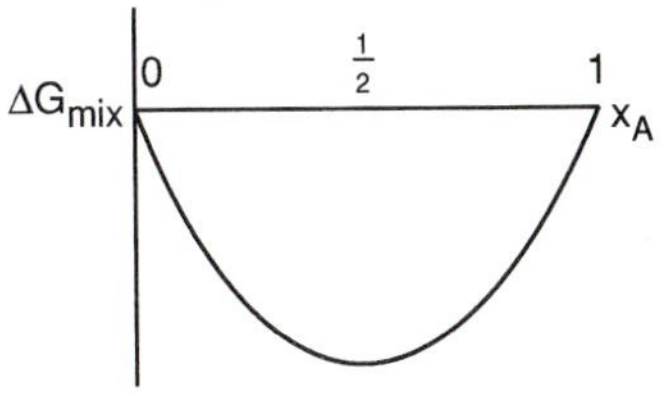

Fig. 2.2 The variation of the Gibbs free energy of mixing with the mole fraction of species A present.

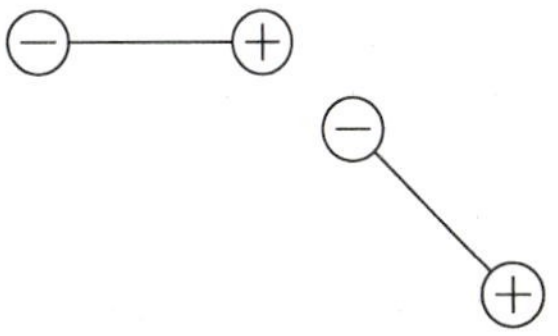

Intermolecular forces are required to keep the molecules in the liquid state. The forces may arise, as above, through the attraction of two dipoles (as in liquid HF) or through van der Waals forces between the molecules (as in liquid benzene, C_6H_6).

The fact that $\Delta H_{mix} = 0$ for an ideal solution means that no heat is given out or taken in when its two components, A and B, are mixed. If the forces between the A and B molecules were stronger than between the A molecules and between the B molecules separately then heat would be evolved on mixing. If they were weaker heat would be absorbed.

electrodes and their potentials we ought, when using the Nernst equation, to use activities rather than concentrations. To do this it is necessary to calculate γ so as to relate these two quantities through eqn (2.4).

2.3 The interpretation of activity coefficients

Let us consider non-ideal solutions of electrolytes, such as shown in the lower container in the diagram below. The solution will contain equal

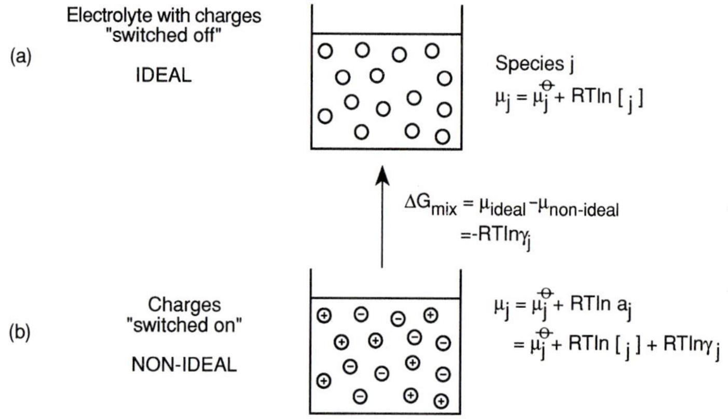

Fig. 2.3 An electrolyte solution. (a) An imaginary situation with the ionic charges absent, (b) The actual non-ideal situation with charges present.

numbers of cations and anions. The disparate interactions between the various chemical species—cations, anions and solvent—lead us to contemplate the solution as non ideal as discussed in the previous section. It follows that for any ion, j, the chemical potential will be given by

$$\mu_j = \mu_j^\ominus + RT\ln a_j \tag{2.15}$$

or,

$$\mu_j = \mu_j^\ominus + RT\ln[j] + RT\ln\gamma_j \tag{2.16}$$

Suppose we now *imagine* exactly the same solution but with the ionic charges absent or switched off. This is depicted in the upper part of the figure above. Under these conditions the solution will be very much more close to ideal behaviour so that

$$\mu_j = \mu_j^\ominus + RT\ln[j] \tag{2.17}$$

It is interesting to question the molar free energy change of the ion j in switching the charges off. That is to say, what is the value of ΔG corresponding to going from the lower container to the hypothetical top container? This will be given by the difference in the chemical potential between the two cases:

$$\Delta G = (\mu_j^\ominus + RT\ln[j]) - (\mu_j^\ominus + RT\ln[j] + RT\ln\gamma_j) = -RT\ln\gamma_j \tag{2.18}$$

It follows that the activity coefficient γ_j in the non-ideal electrolyte solution is intimately related to the free energy change on de-charging the ions. If γ_j is less than unity the 'ion' has a lower free energy in the charged state whereas if γ_j is greater than unity it would be more stable in the imaginary uncharged condition. For dilute electrolyte solutions experiment, as we will see in Chapter 4, tells us that $\gamma_j < 1$. This corresponds to the solution of ions being more stabilised by the presence of charge as compared to its absence.

The question arises next as to why the charge should help stabilise the solution? At first sight this appears paradoxical since, as there are equal numbers of cations and anions in the solution we might reasonably expect as much destabilisation through repulsions between like ions as there is stabilisation between unlike ions.

The cause of the stabilisation can be understood if we consider one single ion in an aqueous sodium chloride solution—say a cation—and think about its motion as it moves about. The solution contains cations, anions and solvent molecules. The latter are present in a much greater quantity. For example the concentration of water in pure water is

$$[H_2O] = \frac{\text{Mass of one litre of water}}{\text{Relative Molecular Mass of } H_2O} \quad \text{(at } 25°\text{C)}$$
$$= \frac{1000}{18} \approx 56 \text{ M}$$

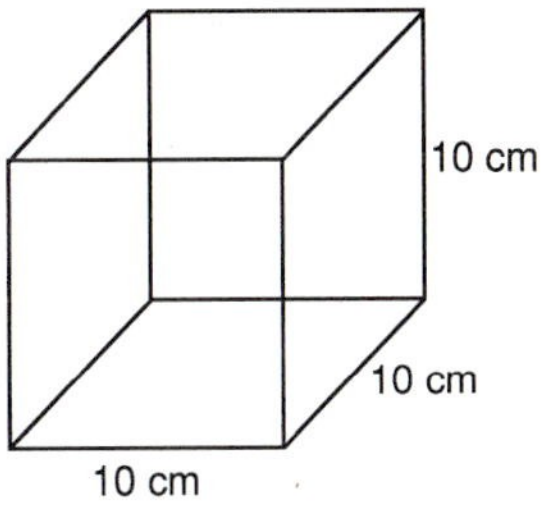

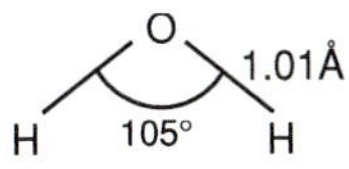

Consider a 10 cm × 10 cm × 10 cm cube of a 0.1 M solution of NaCl.
This contains a litre of solution so that there are $0.1 \times 6 \times 10^{23}$ Na^+ ions together with a similar number of Cl^- ions. If the mean $Na^+ - Na^+$ distance is d cm then,

$$\left(\frac{10}{d}\right)^3 = 0.1 \times 6 \times 10^{23} = 60 \times 10^{21}$$

so that $d \approx 2.5 \times 10^{-7}$ cm = 2.5 nm = 25 Å. This is a large distance compared to the size of a water molecule.

So in a 0.1 M solution of sodium chloride there are over 500 water molecules for each sodium cation or chloride anion. It follows that the average ion–ion separation is quite large. Thus as our cation moves, it predominantly encounters solvent molecules for the overwhelmingly large part of its travels. However from time to time it will meet other ions, both cations and anions. If the motion of all the ions in the solution were entirely random then our cation would encounter other cations as often as it encountered other anions so that the ion would be destabilised through cationic repulsion as often as it is stabilised by cationic/anionic attraction. However some thought shows that *because the ions either attract or repel one another the chances are that the cation will encounter slightly more anions than cations on its travels*. In other words the motion of the cation—or that of any other ion—through the solution is largely but not entirely random; the ionic motions are such that oppositely charged ions encounter each other with a slightly greater frequency than like-charged ions. This has the effect that there is a net stabilisation of the ions in comparison with either the random motion situation, or the hypothetical case where the ions are uncharged.

It was derived above that the quantity $RT\ln\gamma$ represents the Gibbs free energy difference in a non-ideal solution between the case where the charge of the ions is switched off (approximating then to ideality) or switched on. Our discussion in this section of the ion motion leads us to recognise that the physical basis for this free energy difference is, at least in dilute solutions, the coulombically semi-organised motion of the ions which ensures that like charged ions encounter each other less than oppositely charged ions, so conferring a stabilisation on the solution.

We will see later that in concentrated solutions attention must also be paid to ion–solvent interactions. This will be discussed in Section 2.5.

2.4 Debye–Hückel theory

We have noted that for a dilute electrolyte solution the activity coefficient, γ, is usually less than unity, $\gamma < 1$. This implies that the solution is more

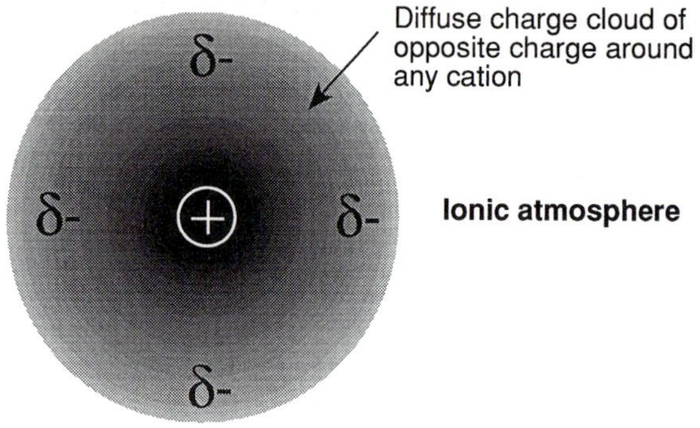

Fig. 2.4 The time average charge cloud around a representative cation.

stable, by an amount $RT\ln\gamma$, as compared to the hypothetical situation where the ionic charge is switched off. The physical origin of this stabilisation was seen in the last section to result from the fact that the anion 'sees' more oppositely charged ions than like-charged ions as it moves about in solution. Let us consider the distribution of charge around an ion. On a *time average* this must be spherically symmetrical and reflect the fact that there will be a build up of opposite charge around the ion reflecting the other ions 'seen'. Figure 2.4 shows a representation of the time average charge distribution around a representative cation. This takes the form of a diffuse spherical cloud of negative charge. The magnitude of the negative charge decreases in a radial direction away from the cation. Far away from the ion the charge becomes zero corresponding to bulk solution, sufficiently remote from the cation that its electroneutrality is unperturbed by the latter. The charge distribution for a typical anion is shown in Fig. 2.5. The spherically symmetrical charge clouds shown in Figs 2.4 and 2.5 are referred to as the *ionic atmospheres* of either the cation or anion in question. The net charge on the atmosphere in each case is a unit charge of sign opposite to that on the central ion.

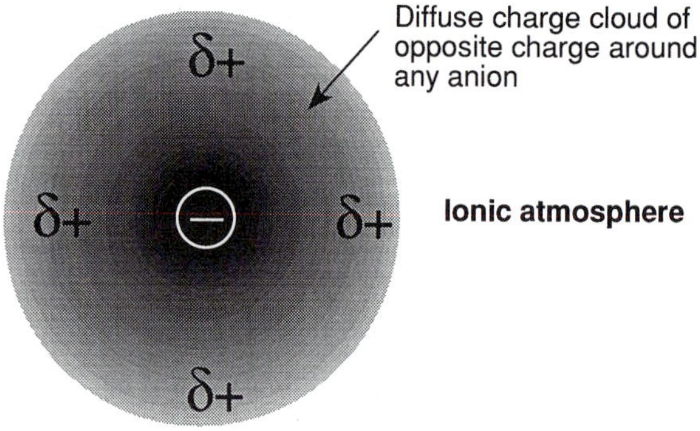

Fig. 2.5 The time average charge cloud around a representative anion.

The insights given by Figs 2.4 and 2.5 provide us with a means for calculating the activity coefficients for both the cations and anions in the solution. In particular, we can calculate the charge distribution in the ionic atmosphere around a particular ion, j, and then use this to quantify the stabilisation of the ion. When scaled up for one mole of ions this should be equal to $RT\ln\gamma_j$ where γ_j is the activity coefficient of ion j. It transpires that this exercise—'Debye–Hückel Theory'—is a straightforward but tedious exercise in electrostatics provided some assumptions (see below) are made. The result is quite simple. The deviation from ideality depends on a quantity known as the ionic strength, I, of the solution. This is defined as

A full mathematical derivation of Debye–Hückel theory is given in Chapter 2 of *Electrochemistry* by P.H. Rieger, published by Prentice–Hall, New Jersey (1987), to which the mathematically inclined reader is directed.

$$I = \frac{1}{2}\sum_i c_i z_i^2 \tag{2.19}$$

where the sum is over all the ions, i, in the solution, c_i is the concentration of ion i and z_i is its charge. The use of eqn (2.19) is exemplified in Box 2.1.

Box 2.1 Calculating ionic strengths

The ionic strength, I, is defined as

$$I = \tfrac{1}{2}\,[\text{sum over every ion of the product of its concentration and the square of its charge}]$$
$$= \tfrac{1}{2}\sum_i c_i z_i^2$$

For example consider a 0.1 M solution of $MgCl_2$:

$$I = \tfrac{1}{2}\,[0.1 \times (+2)^2 + 0.2 \times (-1)^2]$$
$$= 0.3\ \text{M}$$

The ionic strength here is greater than the concentration. As a second example consider a 0.1 M solution of NaCl.

$$I = \tfrac{1}{2}\,[0.1 \times (+1)^2 + 0.1 \times (-1)^2]$$
$$= 0.1\ \text{M}$$

In this case the ionic strength equals the concentration. This is a general result for species of the formula M^+X^-.

The basic equation of Debye–Hückel theory is

$$\log_{10}\gamma_j = -Az_j^2\sqrt{I} \tag{2.20}$$

where z_j is the charge on the ion and A is a temperature and solvent dependent parameter. For water at 25°C, $A \approx 0.5$. In calculating the electrostatic stabilisation conferred on an ion by its atmosphere so as to establish eqn (2.20) several assumptions are made:

i) The cause of the solution non-ideality resides exclusively in coulombic interactions between the ions, and not at all, for example, in ion–solvent interactions.

Box 2.2 The force between ions

The force between two ions carrying charges z_1, and z_2 a distance r apart is given by Coulomb's law:

$$\text{Force} = \frac{z_1 z_2}{4\pi\epsilon_0\epsilon_r} \cdot \frac{1}{r^2}$$

where ϵ_0 is the *vacuum permittivity* and has a value of $8.854 \times 10^{-12}\ C^2\ J^{-1}\ m^{-1}$. ϵ_r is the *relative permittivity*, or *dielectric constant*, of the medium separating the ions. The energy corresponding to the above is

$$\text{Energy} = -\left(\frac{z_1 z_2}{4\pi\epsilon_0\epsilon_r}\right) \cdot \frac{1}{r}$$

This is the energy required to move the ions to a separation, r, from a distance where they are infinitely separated. This quantity is shown graphically below for two ions either of like or unlike charge.

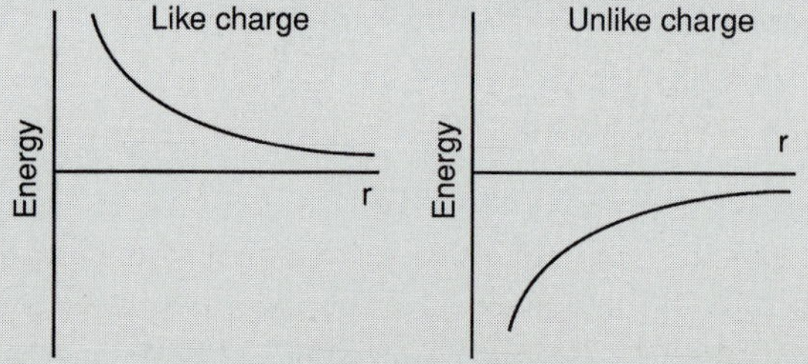

The quantity ϵ_r measures the extent to which a solvent can reduce the energy of interaction of ions dissolved in it. Some values of ϵ_r for common solvents at 25 °C are

C_6H_6	2
$(C_2H_5)_2O$	4
Pyridine	12
CH_3OH	33
CH_3CN	36
Me_2SO	47
H_2O	78

The value of ϵ_r loosely correlates with the polarity of the solvent since dipolar solvent molecules can be partially aligned by the ions:

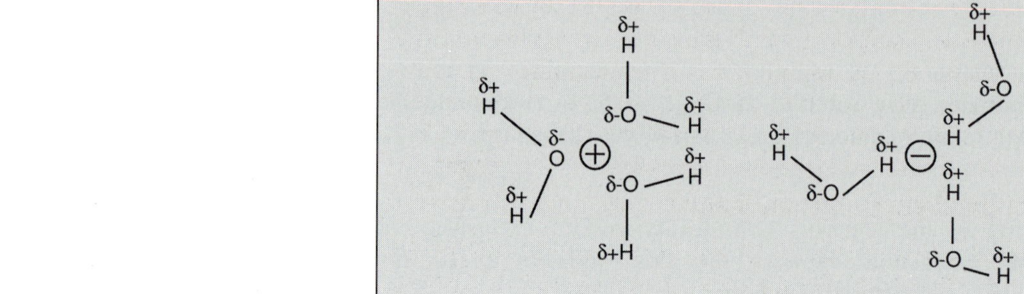

ii) The ionic interactions are quantitatively described by Coulomb's law for point charges (Box 2.2). This presumes that the effects of the solvent is solely to reduce inter-ionic forces by means of the dielectric constant.
iii) The electrolyte is fully dissociated and no significant numbers of ion pairs exist. This implies that the external forces between the ions are weaker than the thermal forces in the solution moving ions around, together and apart.

Ion pairs are formed when a cation and an anion are found in solution as a single entity:

⊖⊕

They often exist when electrolytes are dissolved in solvents of low polarity.

These assumptions work well in dilute solutions, so that for ionic concentrations below approximately 10^{-2} M, equation 2.20, known as the *Debye–Hückel limiting law* works quantitatively.

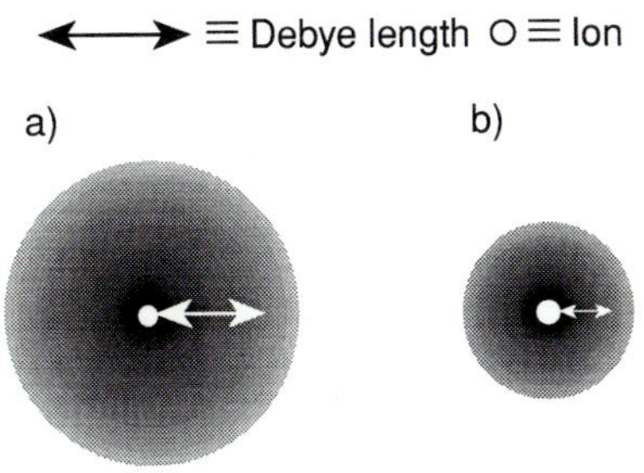

Fig. 2.6 The relative size of the ionic atmospheres for aqueous solutions containing a simple M^+X^- electrolyte at (a) low, and (b) high ionic strength.

The Debye–Hückel limiting law predicts that the deviation from ideality increases with the square root of the ionic strength, *I*. It is interesting to consider why this should be, and to focus on the size of the ionic atmosphere. The effective size of the latter is measured by its *Debye length*, which gives an indication of the distance between any ion and the *average* location of the charge in its ionic atmosphere as illustrated in Fig. 2.6. The higher the concentration the shorter the Debye length. For example in a 10^{-3} M aqueous solution of electrolyte of formula MX the Debye length is close to 100 Å; for a 10^{-1} M solution it is 10 Å. In general for such electrolytes,

$$\text{Debye length} \propto 1/\sqrt{I} \tag{2.21}$$

It follows that as the ionic strength increases the distance between the central ion and the charge in the ionic atmosphere shrinks. Accordingly Coulomb's law (Box 2.2) leads us to expect that the electrostatic stabilisation of the ion conferred by the ionic atmosphere increases so that γ_j becomes smaller and the solution more non-ideal.

Last, notice that the size of the ionic atmosphere is typically of the order of 10 Å–100 Å. This is very much *larger* than the size of the cations or anions in solution. This again emphasises first, the large scale of the mean ion separation in dilute electrolyte solutions and second, that ion-pairs are *not* considered as sources of non-ideality within the Debye–Hückel theory.

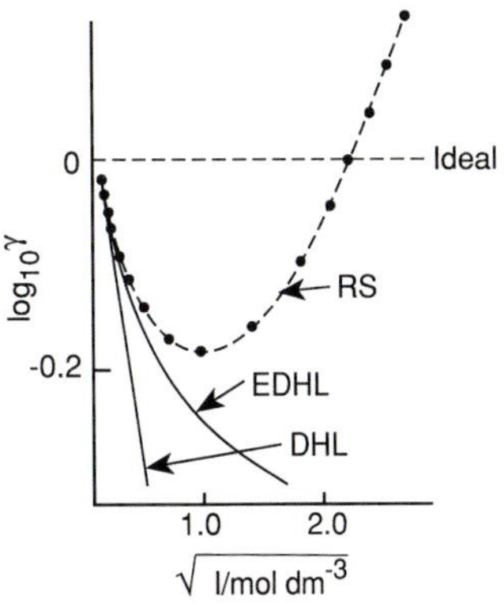

Fig. 2.7 Experimental activity coefficients (•) and their ionic strength dependence shown schematically. DHL ≡ Debye–Hückel limiting law; EDHL ≡ Extended Debye–Hückel law; RS ≡ Robinson and Stokes equation.

2.5 Limits of the Debye–Hückel theory

Chapter 4 will describe how activity coefficients may be measured. Typical results are gathered in Fig. 2.7 Notice the *y* axis gives the logarithm (to base

10) of the ionic activity coefficient γ_j; $\log_{10}\gamma_j = 0$ corresponds to $\gamma_j = 1$ and so to ideality. The x axis displays the square root of the ionic strength. The experimental data are represented by the points. It can be seen that the Debye–Hückel limiting law (labelled DHL), $\log_{10}\gamma_j = -Az^2_j\sqrt{I}$, works well in dilute solution—up to concentrations around 10^{-2} M—but overestimates the deviation from ideality at higher concentrations.

If the ion radius, a, is put equal to zero in the extended Debye–Hückel law the equation representing the simple Debye–Hückel limiting law is regained.

The line labelled EDHL represents the equation

$$\log_{10}\gamma_j = \frac{-Az_j^2\sqrt{I}}{1 + Ba\sqrt{I}} \qquad (2.22)$$

which is the extended Debye–Hückel law. It is very similar to eqn (2.20) except that new terms appear in the denominator. The constant B is, like A, a solvent and temperature specific parameter, whilst a is the radius of the ion, j. Equation (2.22) is established in exactly the same way as eqn (2.20) except the the ions are treated not as point charges but as spheres of radius a. Fig. 2.7 shows that the extended Debye–Hückel law applies quantitatively up to slightly higher concentrations than the Debye–Hückel limiting law.

Neither eqn (2.20) or (2.22) can however predict the upturn in the values of γ seen in Fig. 2.7. Both for the Debye–Hückel limiting law and the extended Debye–Hückel law increasing the ionic strength causes $\log\gamma_j$ to decrease even further below zero. Physically this is because both equations attribute the deviation from ideality to electrostatic forces stabilising each ion and these increase—and the ionic atmospheres shrink—as the ionic strength increases. Comparing theory and experiment in Fig. 2.7 suggests that some new factor must become important at higher concentrations.

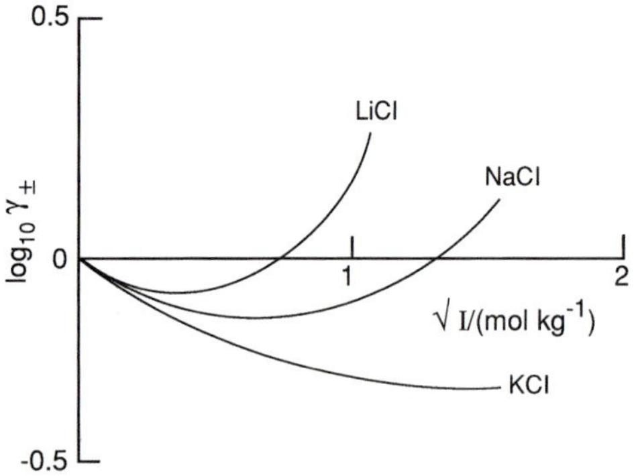

Fig. 2.8 The variation of the activity coefficients $\gamma_\pm$ of LiCl, NaCl and KCl with the ionic strength of the solution.

A hint as to the identity of this new factor emerges if we compare the activity coefficients of aqueous solutions of LiCl, NaCl and KCl, as in Fig. 2.8. The y axis no longer relates to the activity coefficient of a single ion, but is the *mean activity coefficient*, $\gamma_\pm$, defined as

$$\gamma_\pm = (\gamma_+\gamma_-)^{1/2} \qquad (2.23)$$

for an electrolyte of stoichiometry MX.

More generally the mean activity coefficient for a salt of stoichiometry, M_aX_b, is given by

$$\gamma_\pm = (\gamma_+^a\gamma_-^b)^{1/(a+b)}$$

The corresponding form of the Debye–Hückel limiting law for $\gamma_\pm$ is

$$\log_{10}\gamma_\pm = -A \mid z_+z_- \mid \sqrt{I}$$

where z_+ and z_- are the charges on the cation and anion.

It can be seen that the deviation of experiment from the Debye–Hückel limiting law and the extended Debye–Hückel law as the ionic strength is increased is soonest for LiCl and latest for KCl. This implies that the new factor influencing deviation from ideality is greatest for Li^+ and least for K^+.

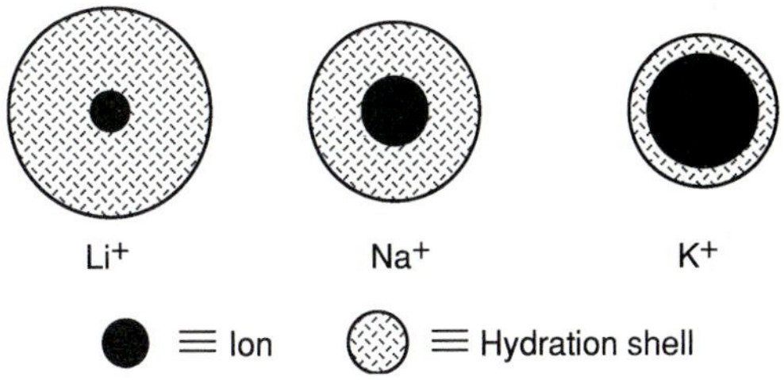

Fig. 2.9 The hydration of lithium, sodium and potassium ions.

The charge density on the alkali metal cations follows the sequence $Li^+ > Na^+ > K^+$. As a consequence Li^+ cations are the most strongly hydrated in aqueous solution (as evidenced in Table 2.1) and K^+ cations the least as shown in Fig. 2.9. This observation suggests, as is the case, *that the deviation from the Debye–Hückel limiting law and the extended Debye–Hückel law at higher concentrations is due to ion–solvent effects.*

Table 2.1: The Hydration of the ions Li^+, Na^+ and K^+

Species	Li^+	Na^+	K^+
Crystal radii[a]/ Å	0.60	0.95	1.33
Hydrated Radii[b] / Å	2.4	1.8	1.3
Hydration Numbers[c]	2-22	2-13	1-6

[a]From x-ray diffraction
[b]From Stokes' Law; see Chapter 3
[c]Estimated via many different techniques. The wide ranges reflect the ill-defined nature of the hydration shells and the multiplicity of experimental approaches.

The way in which ion–solvent effects can influence activity coefficients can be understood in the light of the following argument. We have seen (Section 2.3) that the quantity $RT\ln\gamma_j$ is the free energy of stabilisation of ion j conferred by virtue of its charge. Values of $\gamma_j < 1$ correspond to negative (favourable) free energies of this type whereas $\gamma_j > 1$ correspond to positive values in which the ion is destabilised in its charged state as compared to the hypothetical uncharged condition. Inspection of Figs 2.7 and 2.8 shows that at high ionic strengths (concentrations) γ_j rises above unity corresponding to a positive value of $RT\ln\gamma_j$, and so the ion is relatively destabilised when charged as compared to uncharged. However the concentrations at which this occurs are rather high—of the order of one molar or more. The ions of electrolytes are hydrated when dissolved in water and the number of water molecules associated with each ion can be quite high. Typically about six water molecules are attached *directly* to a cation but rather more are indirectly attached through hydrogen

Ionic hydration is crucial to the dissolution of salts in water since the energy released by hydration is used to offset that required to break up the crystal—the lattice energy.

bonding outside the primary solvation shell as shown in Fig. 2.10. For example in the case of Na^+ cations, estimates of the total number of molecules of hydration range from 2 to 13.

Details of hydration numbers and how to measure them are given in J. Burgess, *Metal Ions in Solution*, Ellis Horwood, 1978. Note that different techniques give contrasting answers since they measure different physical properties and since the outer boundary of the second hydration shell is necessarily rather indistinct.

Supposing we take a value of, say 10, together with a similar value for the chloride anion, Cl^-, we can see than in a 1 M solution of NaCl approximately 20 moles out of every 56 moles of water will be 'locked up' as water of hydration of the ions. The source of ionic destabilisation at higher concentrations can then be understood due to the ever-reducing availability of water to fully solvate the ions. That is, in concentrated aqueous solutions, due to the finite amounts of water available, the solvation of the ions becomes increasing imperfect and they become relatively destabilised. This is reflected in the trends in γ_j and $\gamma_\pm$ shown in Figs 2.7 and 2.8.

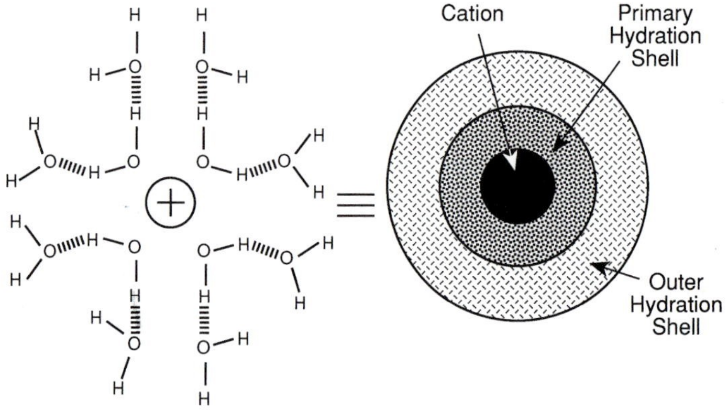

Fig. 2.10 Primary and outer hydration shells.

The effects of hydration—ion–solvent interactions—on activity coefficients can be quantified by means of the Robinson and Stokes Equation:

$$\log_{10}\gamma_\pm = -\left(\frac{A(z_+z_-)\sqrt{I}}{1 + aB\sqrt{I}}\right) + cI \tag{2.24}$$

where c is treated as a solute and solvent specific parameter characterising the solvation of the ions. Examination of Fig. 2.7 shows that eqn (2.24) labelled RS is able to describe experimental activity coefficient data quantitatively over the full concentration range. The first term on the right hand side of eqn (2.24) is the extended Debye–Hückel expression for ion–ion interactions. The second term on the right hand side accounts for ion–solvent interactions and the loss of ion stability at high ionic strengths. Note the two quantities in eqn (2.24) have opposite signs.

2.6 Applications of the Debye-Hückel limiting law

In this section we consider two applications of the Debye–Hückel limiting law. One is to the solubility of poorly ('sparingly') soluble salts and the other is to reaction kinetics.

Solubilities

The solubility of sparingly soluble salts can be slightly enhanced by an increase in ionic strength.

For example the solubility product of silver chloride is $K_{sp} = a_{Ag^+}.a_{Cl^-} = 1.7 \times 10^{-10}$ mol^2 dm^{-6} so that in pure water the solubility is approximately 1.3×10^{-5} mol dm^{-3}. The solubility of AgCl is promoted, *a little*, by the addition of KNO_3. This is known as 'salting in'. The increase in solubility can be understood for a general case by considering the following process:

$$M^{z+}X^{z-}(s) \rightleftharpoons M^{z+}(aq) + X^{z-}(aq)$$

The equilibrium constant for this dissolution is given by

$$K_{sp} = a_{M}.\, a_{X} = \gamma_M[M^{z+}].\gamma_X[X^{z-}] \quad (2.25)$$

$$= \gamma_\pm^2 [M^{z+}]\, [X^{z-}] \quad (2.26)$$

The cation and anion concentrations will be equal to each other:

$$[M^{z+}] = [X^{z-}] = c \quad (2.27)$$

so that

$$\log_{10}K_{sp} = 2\log_{10}\gamma_\pm + 2\log_{10}c \quad (2.28)$$

Applying the Debye–Hückel limiting law then gives

$$\log_{10}K_{sp} = -2Az^2\sqrt{I} + 2\log_{10}c \quad (2.29)$$

which, when rearranged, shows the variation of the concentration of the dissolved ions with the ionic strength, I:

$$\log_{10}c = \tfrac{1}{2}\log_{10}K_{sp} + z^2A\sqrt{I} \quad (2.30)$$

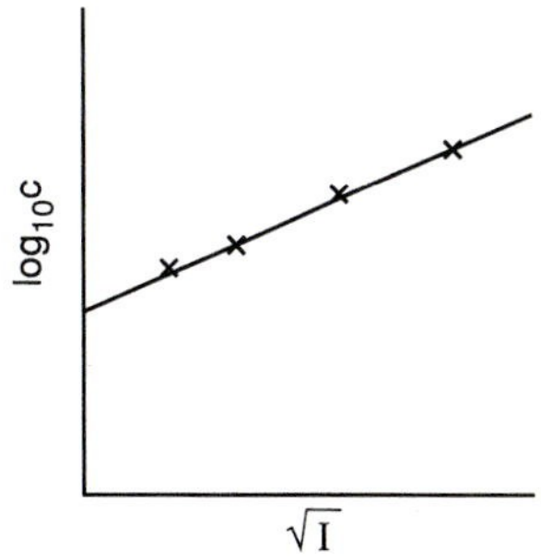

Fig. 2.11 The salting in of a sparingly soluble salt as the ionic strength of the solution increases.

This equation is illustrated in Fig. 2.11. The figure shows that as the ionic strength is increased the solubility of MX is promoted. It is helpful to ask the physical reason for this and to focus on the specific case of silver chloride. When AgCl is dissolved into a solution to which KNO_3 is progressively added, the Ag^+ and Cl^- ions will develop ionic atmospheres around themselves which will serve to stabilise the ions. Considering the equilibrium

$$AgCl\,(s) \rightleftharpoons Ag^+\,(aq) + Cl^-\,(aq)$$

stabilisation will 'pull' the equilibrium to the right hand side, so promoting the solubility of the silver halide. As the ionic strength of the solute rises the Debye length of the atmosphere will progressively shrink so that each ion will become closer and closer to the (opposite) charge surrounding it and so the stabilisation will be enhanced. The solubility will be further increased. This is shown in Fig. 2.12.

In deriving 2.30 we used the Debye–Hückel limiting law. This is, itself, only valid for concentrations below about 10^{-2} M so that the final result will equally only apply quantitatively to appropriately dilute solutions.

Last we note that ionic strength effects are not necessarily the only ideas required to understand how the composition of a solution may change if

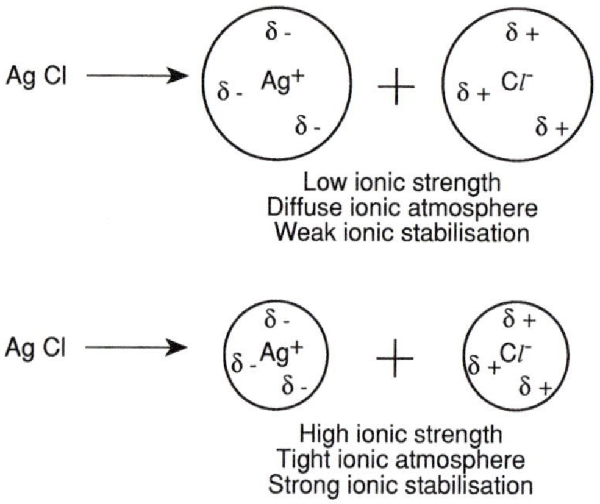

Fig. 2.12 The dissolution of AgCl.

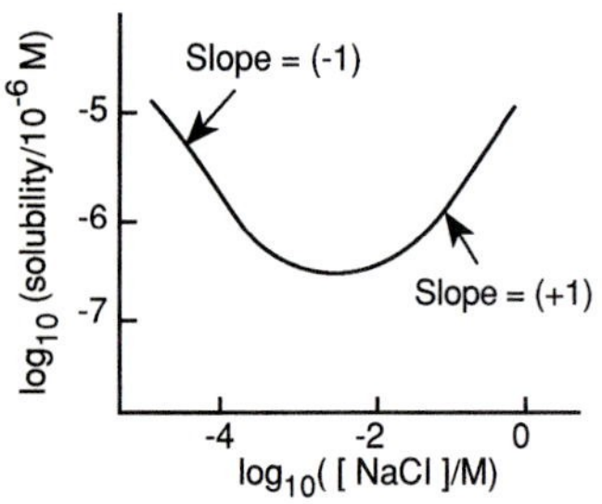

Fig. 2.13 The solubility of AgCl in aqueous KCl.

electrolyte is added to it. In particular it should be noted that the *common ion effect* and *complexation reactions* are frequently important. These can be illustrated if we return to our example of a saturated solution of silver chloride and consider now the addition of KCl rather than KNO_3. The solubility as measured by the total concentration of dissolved Ag^+, varies with the added chloride ion concentrations as shown in Fig. 2.13. On the left hand side of the figure the solubility decreases as the chloride ion concentration increases. This is the common ion effect and can be understood on the basis of the equilibrium,

$$AgCl\ (s) \rightleftharpoons Ag^+\ (aq) + Cl^-\ (aq)$$

so that, neglecting activity effects

$$K_{sp} = [Ag^+][Cl^-] \tag{2.31}$$

This implies that $[Ag^+]$ decreases as $[Cl^-]$ increases. Mathematically,

$$\log[Ag^+] = \log K_{sp} - \log[Cl^-] \tag{2.32}$$

which predicts that the slope of Fig. 2.13 should be (–1) in the region where the common ion effect operates. This is seen to be the case.

At high chloride concentration, on the right hand side of Fig. 2.13 the solubility rises as the chloride ion concentration rises. This is as a result of complexation and the formation of the linear species $AgCl_2^-$:

$$AgCl(s) + Cl^-(aq) \rightleftharpoons AgCl_2^-(aq)$$

If the equilibrium constant for this reaction is K then, again neglecting any activity effects,

$$K = \frac{[AgCl_2^-]}{[Cl^-]}$$

This predicts that the amount of $AgCl_2^-$ and hence that of the dissolved silver will now rise as $[Cl^-]$ increases. Mathematically,

$$\log[AgCl_2^-] = \log K + \log[Cl^-] \tag{2.33}$$

This shows that where the complexation process occurs the slope of Fig. 2.17 should be (+1) as is seen to be the case.

The kinetic salt effect

Consider the reaction between two charged species, M and X:

$$\mathrm{M} + \mathrm{X} \rightleftharpoons \{\mathrm{M,X}\} \xrightarrow{k} \text{products}$$

where {M,X} is an activated complex or transition state denoted $\ddagger$ in the following text. If we assume the transition state forms a pre-equilibrium with the reactants M and X prior to reaction of the former with a rate constant k (s^{-1}) then

$$K = \frac{a_\ddagger}{a_\mathrm{M} a_\mathrm{X}} = \frac{\gamma_\ddagger}{\gamma_\mathrm{M}\gamma_\mathrm{X}} \frac{[\ddagger]}{[\mathrm{M}][\mathrm{X}]} \tag{2.34}$$

The rate of reaction will be given by

$$\text{Rate} = k[\ddagger] \tag{2.35}$$

Combining (2.34) and (2.35) gives

$$\text{Rate} = kK\frac{\gamma_\mathrm{M}\gamma_\mathrm{X}}{\gamma_\ddagger}[\mathrm{M}][\mathrm{X}] \tag{2.36}$$

Therefore M and X react with an apparent second order rate constant given by

$$k_\mathrm{app} = kK\frac{\gamma_\mathrm{M}\gamma_\mathrm{X}}{\gamma_\ddagger} \tag{2.37}$$

or

$$\log_{10}k_\mathrm{app} = \log_{10}kK + \log_{10}\gamma_\mathrm{M} + \log_{10}\gamma_\mathrm{X} - \log_{10}\gamma_\ddagger \tag{2.38}$$

When expanded taking into account the Debye–Hückel limiting law this gives

$$\begin{aligned}\log_{10}k_\mathrm{app} &= \log_{10}kK - Az_\mathrm{M}^2\sqrt{I} - Az_\mathrm{X}^2\sqrt{I} + A(z_\mathrm{M}+z_\mathrm{X})^2\sqrt{I} \\ &= \log_{10}kK + 2Az_\mathrm{M}z_\mathrm{X}\sqrt{I}\end{aligned} \tag{2.39}$$

or

$$\log_{10}\left(\frac{k_\mathrm{app}}{k_{I\to 0}}\right) = 2Az_\mathrm{M}z_\mathrm{X}\sqrt{I} \tag{2.40}$$

where $k_{\mathrm{I}\to 0}$ is the measured second order rate constant at infinite dilution, and z_M and z_X are the charges on the ions M and X respectively.

Equation (2.40) is an interesting result. It predicts that if we change the ionic strength, I, of a solution by adding an inert electrolyte containing no M or X, and which plays no part in the reaction other than to change the ionic strength, nevertheless the rate of the reaction between M and X can be altered. Intriguingly if M and X have the same charge (both positive, or both negative) increasing the ionic strength is predicted to increase the reaction rate by eqn (2.40). In contrast if they have opposite charges then the rate is anticipated to decrease!

The clue to this behaviour lies in the effect of ionic atmospheres, not just on the reactants M and X, but also now on the transition state, $\ddagger$. Imagine a divalent cation and a divalent anion reacting:

$$\mathrm{M}^{2+} + \mathrm{X}^{2-} \rightleftharpoons \{\ddagger\}^0 \rightarrow \text{products}$$

Suppose we add an inert salt. The effect of this will be to supply ions which provide the reactants with an ionic atmosphere. This will stabilise the ions. In contrast the transition state, which is neutral, will have no

The rates of reactions in solution are discussed in detail in the Oxford Chemistry Primer entitled *Modern Liquid Phase Kinetics* written by B. G. Cox (OCP 21).

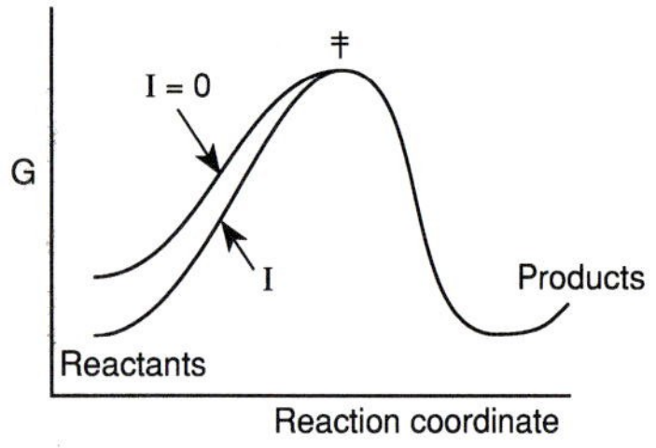

Fig. 2.14 Reaction free energy profile for the reaction of a (2+) cation and a (2–) anion (a) at zero ionic strength, and (b) at a finite ionic strength, *I*.

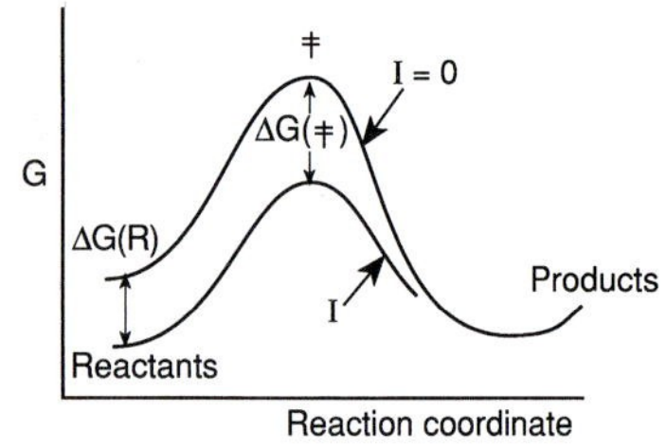

Fig. 2.15 Reaction free energy profile for the reaction of two (2+) cations, (a) at zero ionic strength, and (b) at a finite ionic strength, *I*.

ionic atmosphere and hence its energy will be essentially unchanged. The effect on the free energy profile of the reactants on going from infinite dilution to an ionic strength of *I* is shown in Fig. 2.14. It can be seen that the barrier to the reaction has been increased so that its rate must slow down as predicted by eqn (2.40).

Next consider two divalent cations reacting:

$$M^{2+} + X^{2+} \rightleftharpoons \{\ddagger\}^{4+} \rightarrow \text{Products}$$

and note that now the transition state carries a charge (4+). The free energy profile, analogous to the one in Fig. 2.14, is shown in Fig. 2.15. In this case both the reactants (R) and the transition state (‡) are stabilised, as compared to the case of infinite dilution, if the ionic strength is raised to a value *I*. Consideration of the arguments given in Section 2.3 lead us to be able to quantify the changes in free energy shown in Fig. 2.15:

$$\Delta G(\ddagger) \propto RT\ln\gamma_{\ddagger}$$

$$\Delta G(\mathrm{R}) \propto RT\ln\gamma_{\mathrm{M}} + RT\ln\gamma_{\mathrm{X}}$$

If we assume that the ionic strength is such that the Debye–Hückel limiting law applies, then using the identity

$$\ln\gamma = 2.303\,\log_{10}\gamma$$

we obtain

$$\Delta G(\ddagger) \propto 2.303RT\log_{10}\gamma = -\,(2.303ART)(+4)^2\sqrt{I}$$

$$\Delta G(R) \propto 2.303RT[\log_{10}\gamma_{\mathrm{M}} + \log_{10}\gamma_{\mathrm{X}}] = -\,(2.303ART)[(+2)^2 + (2)^2]\sqrt{I}$$

This shows that the transition state is stabilised by a larger amount than is the reactants since $(+4)^2 > \{(+2)^2 + (+2)^2\}$. It follows that the kinetic barrier to the reaction is decreased and the reaction will speed up as the ionic strength increases.

Figure 2.16 shows the results of experiments conducted on the kinetics of various processes as a function of the ionic strength. It can be seen that the predictions of eqn (2.40) are accurately followed and that the slope of the line for each system is proportional to $(z_{\mathrm{M}}z_{\mathrm{X}})$ as expected. These data can be taken as evidence for the quantitative validity of the Debye–Hückel limiting law for appropriately dilute concentrations.

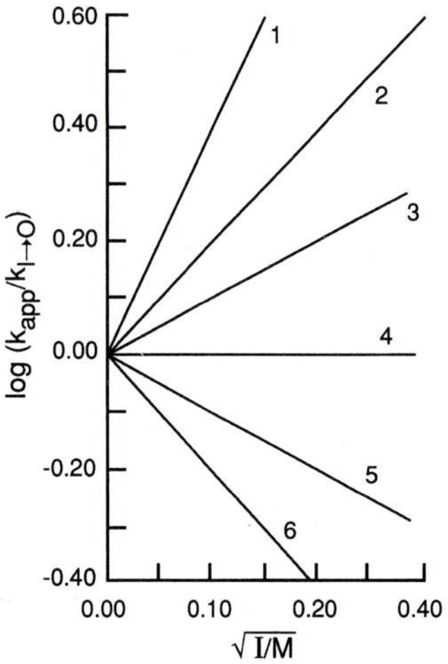

Fig. 2.16 Variations of the rates of some reactions with ionic strength:
(1) $2[Co(NH_3)_5Br]^{2+} + Hg^{2+} + H_2O \rightarrow 2[Co(NH_3)_5H_2O]^{3+} + HgBr_2$ $(z_M z_X = 4)$
(2) $S_2O_8^{2-} + 2I^- \rightarrow I_2 + 2SO_4^-$ $(z_M z_X = 2)$
(3) $OH^- + [NO_2NCOOC_2H_5]^- \rightarrow N_2O + CO_3^{2-} + C_2H_5OH$ $(z_M z_X = 1)$
(4) Inversion of glucose (uncharged) $(z_M z_X = 0)$
(5) $H_2O_2 + 2H^+ + 2Br^- \rightarrow 2H_2O + Br_2$ $(z_M z_X = -1)$
(6) $[Co(NH_3)_5Br]^{2+} + OH^- \rightarrow [Co(NH_3)_5OH]^{2+} + Br^-$ $(z_M z_X = -2)$

3 Migration of ions

In this chapter we will consider only fully dissociated electrolytes. These are known as *strong electrolytes* in contrast with *weak electrolytes* which include, for example, acids such as ethanoic acid, CH_3COOH, which are largely undissociated in aqueous solution. Example of strong electrolytes include aqueous solutions of HCl or KNO_3.

3.1 Current density and voltage gradients

Imagine a solution containing a fully dissociated electrolyte in which are placed two electrodes each of area A. Suppose a current, I, flows between the two electrodes. In the zone of the solution between the electrodes cations will be moving to the negative electrode ('cathode') whilst anions will move towards the positive electrode ('anode'). The movement of the two sorts of ions constitutes the flow of current in the bulk of the solution and it is the factors concerning this process which are of importance in this chapter, although we will briefly consider the nature of the current

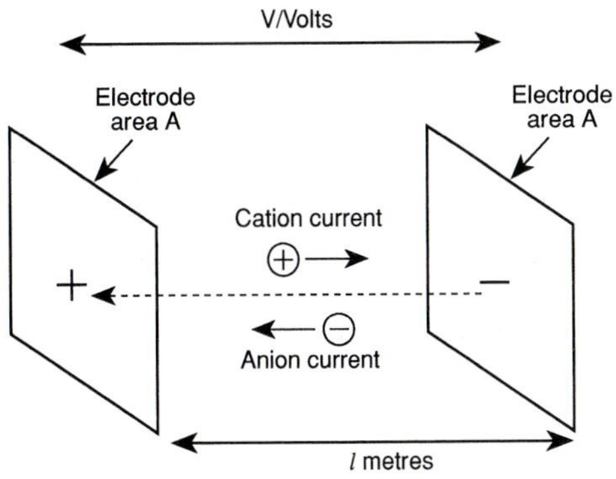

Fig. 3.1 The movement of cations and anions between two electrodes.

flow at electrode/solution interfaces. The experiment of interest is summarised in Fig. 3.1. Note that a voltage, V, is used to drive the current flow in the solution and that the electrodes are a distance, l, apart. The bulk electrolyte obeys the following relationship

$$\kappa = \frac{1}{\rho} = \frac{l}{RA} \tag{3.1}$$

where ρ is the resistivity, κ the *conductivity* ($\Omega^{-1}m^{-1}$) and R is the resistance (Ω) of the solution between the electrodes. The conductivity is a property of the chemical nature and composition of the electrolyte solution. Rearranging eqn 3.1 gives

$$R = \frac{l}{\kappa A} \tag{3.2}$$

which tells us that the resistance to current flow increases the greater the distance between the electrodes and the smaller the electrode area. These predictions are consistent with one's intuitive expectations. It is worth noting that equation (3.2) also describes the current flow in a metal bar where the charge carriers are electrons as well as describing an electrolyte solution where the carriers are cations and anions.

In addition to equation (3.2) the solution also obeys Ohm's Law:

$$V = IR \tag{3.3}$$

This equation may be combined with eqn (3.2) to give the important relationship

$$\kappa = \frac{(I/A)}{(V/l)} \tag{3.4}$$

where the term (I/A) is the *current density* (Amps per unit area, A m^{-2}) and (V/l) is the *voltage gradient* (Volts per metre, V m^{-1}) in the solution. Rearranging eqn (3.4) gives

$$(I/A) = \kappa.(V/l) \tag{3.5}$$

Current density = Conductivity × Voltage Gradient

In eqn (3.5) the voltage gradient can be thought of as the 'driving force' required to induce a current density (I/A) in a medium with a conductivity κ.

3.2 Molar conductivity

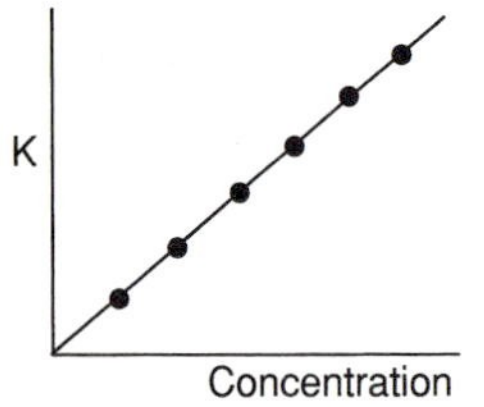

Fig. 3.2 The conductivity is, to a very good approximation, linearly dependent on the concentration of the ions present in solution.

If the conductivity, κ, of various solutions is investigated experimentally it is readily discovered, that to a very high approximation, κ is a linear function of electrolyte concentration. This is illustrated in Fig. 3.2. It is thus helpful to introduce *the molar conductivity*, Λ, defined by

$$\frac{\kappa}{c} = \Lambda \tag{3.6}$$

which has units of Ω^{-1} cm^2 mol^{-1}. The linear dependence of the conductivity on the concentration is readily understood: the more ions to carry the current present the greater is the conductivity for a fixed driving force (voltage gradient). Some values of single ion molar conductivities are shown in Table 3.1.

Examination of the table shows that H^+ and OH^- have anomalously high molar conductivities in aqueous solution. Other trends will be discussed shortly, but note that K^+ and Cl^- have closely similar molar conductivities.

Table 3.1 enables us to predict the molar conductivity, Λ_{salt}, of any salt M_aX_b. On dissolution one mole of the salt will form a moles of cation of

Table 3.1 Single ion conductivities in water (25°C)/Ω^{-1} cm^2 mol^{-1}

Ion	Λ_+	Ion	Λ_-
H^+	350	OH^-	199
Li^+	39	F^-	55
Na^+	50	Cl^-	76
K^+	74	Br^-	78
Ag^+	62	I^-	77
NH_4^+	74	NO_3^-	71
Mg^{2+}	106	SO_4^{2-}	158
Ca^{2+}	119	CH_3COO^-	41
Ba^{2+}	127	$Fe(CN)_6^{4-}$	442

molar conductivity Λ_+ and b moles of anion with molar conductivity Λ_-. Then

$$\Lambda_{\text{salt}} = a\Lambda_+ + b\Lambda_- \qquad (3.7)$$

This is known as the *Law of independent ion migration,* implying that, as drawn in Fig. 3.1, the cations and anions, move essentially independently of each other, to a good approximation.

3.3 Measurement of conductivity and molar conductivity

Conductivity measurements may be readily made using the cell shown in Fig. 3.3. Imagine performing such experiments to examine, say, the conductivity of hydrochloric acid, H^+Cl^-. If the experiment is carried out

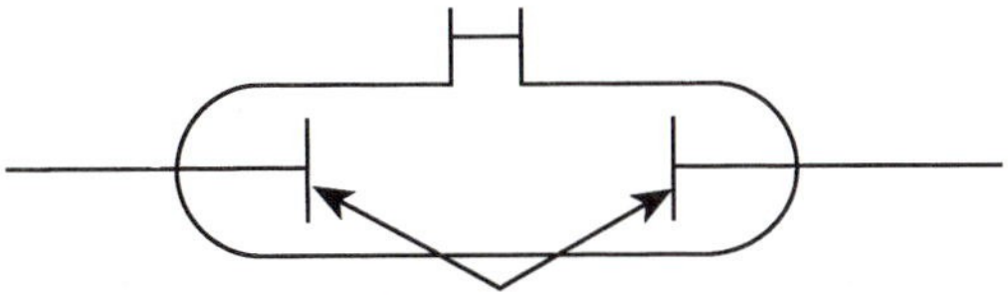

Fig. 3.3 A cell for conductivity measurements.

with a direct current then one electrode will be the anode (+) and the other the cathode (–). The voltage applied between the electrodes will set up a voltage gradient in the solution which will stimulate the migration of cations and anions as described earlier in this chapter. However for a current to be sustained charge transfer across both the electrode/solution interfaces will be required. For this to happen, under direct current conditions, electrolysis must occur. At the anode chlorine will be evolved:

$$Cl^-(aq) \rightarrow \tfrac{1}{2}\,Cl_2(g) + e^-\ (\text{metal})$$

whilst at the cathode, hydrogen will be released

$$H^+(aq) + e^-\ (\text{metal}) \rightarrow \tfrac{1}{2}H_2\ (g)$$

It follows that, as a result of the current flow, the solution will become depleted of ions, and so a result its conductivity will fall. *Consequently it is inappropriate to use direct current to measure conductivities.*

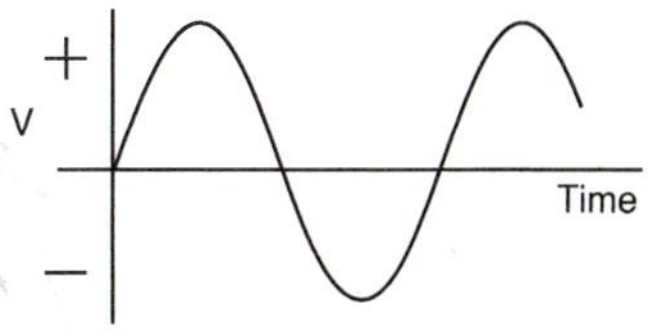

Fig. 3.4 An AC voltage, *V*, is applied between the two electrodes of the cell shown in Figure 3.3 to measure the solution conductivity.

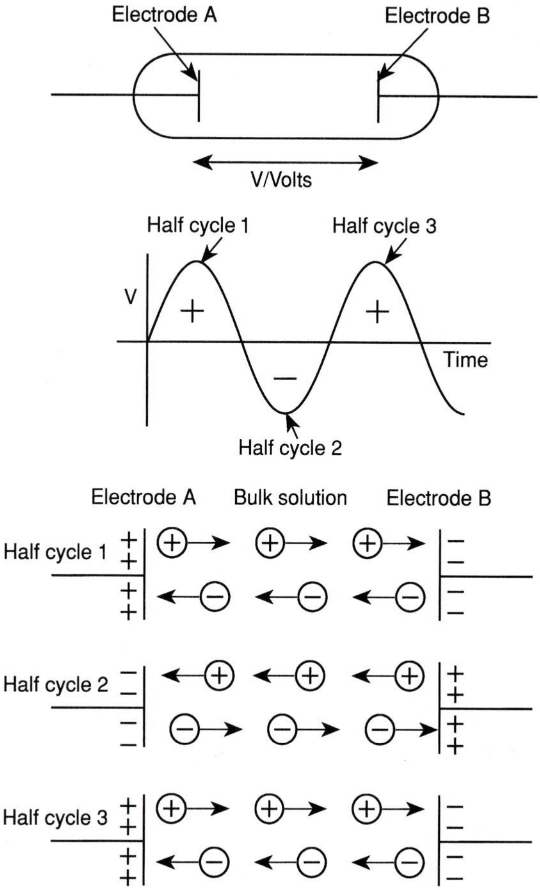

Fig. 3.5 The movement of ions in solution induced by the application of an AC voltage, *V*, between the electrodes A and B of the cell illustrated above.

Almost invariably alternating current is used in conductivity determinations. The frequency employed is generally around 1 kHz so that the polarity of the cell changes 2000 times per second and each electrode takes on the rôle of anode and cathode for half a millisecond at a time. This period of time is too short for electrons to be transferred in or out of solution phase species to or from the electrode so that there is no conversion of chloride to chlorine or of H^+ to hydrogen. The sole rôle of the rapidly fluctuating voltage is to sequentially attract and repel ions from the electrode/solution interface as is illustrated in Fig. 3.5. The movement of ions in and out of the interface constitutes an alternating current in the solution and is thus propagated through the solution between the two electrodes. Conductivity measurements are made by using the cell shown in Fig. 3.3 placed traditionally in the Wheatstone bridge apparatus shown in Fig. 3.6. The oscillator applies an alternating current to the circuit. When the variable resistor R_{var} is adjusted so that no current is passed through the detector (D) then the bridge is 'balanced' and the cell resistance can be found from

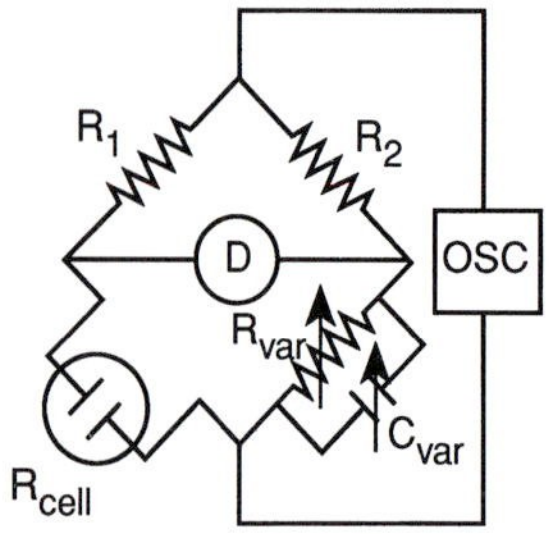

Fig. 3.6 An AC bridge for determining the cell resistance, R_{cell}. R_{var} and C_{var} represent a variable resistance and variable capacitance, respectively.

$$R_{cell} = R_{var}\frac{R_1}{R_2} \tag{3.8}$$

This enables the conductivity, κ, to be found from eqn (3.4) provided a value of (l/A) is known. This is referred to as the cell constant and is most easily found experimentally by using a solution of known conductivity for calibration purposes.

3.4 Transport numbers

Consider an aqueous solution of lithium chloride. Current can be carried in the solution by both lithium cations and chloride anions (Fig. 3.1). The values of the molar conductivities of the two ions are $\Lambda_{\mathrm{Li}} = 38.7\ \Omega^{-1}\ \mathrm{cm}^2\ \mathrm{mol}^{-1}$ and $\Lambda_{\mathrm{Cl}} = 76.3\ \Omega^{-1}\ \mathrm{cm}^2\ \mathrm{mol}^{-1}$. From this it is evident that the majority of the current is carried by the chloride ion. It is helpful to introduce the concept of the transport numbers, t_+ and t_-, which describe the fraction of the current carried by the cation and the anion respectively. They are defined by

$$t_+ = \frac{\Lambda_+}{\Lambda_+ + \Lambda_-} \quad \text{and} \quad t_- = \frac{\Lambda_-}{\Lambda_+ + \Lambda_-} \tag{3.9}$$

For the case of aqueous lithium chloride;

$$t_+ = \frac{38.7}{38.7 + 76.3} = 0.34 \quad \text{and} \quad t_- = \frac{76.3}{38.7 + 76.3} = 0.66$$

and it is evident that 66% of the current is carried by the chloride ion. In contrast for a solution of potassium chloride

$$t_+ = \frac{73.5}{73.5 + 76.3} = 0.49 \quad \text{and} \quad t_- = \frac{76.3}{73.5 + 76.3} = 0.51$$

and the two ions effectively carry the current equally between them.

3.5 A simple model for single ion molar conductivities

Figure 3.7 shows a representation of the forces acting upon a cation of charge z in a voltage gradient of V/l, The ion will feel an electrical attraction to the negative electrode by virtue of its charge. The electrical *force* of attraction is

$$\text{Electrical Force} = zeV/l \tag{3.10}$$

where e (coulombs) is the charge on an electron.

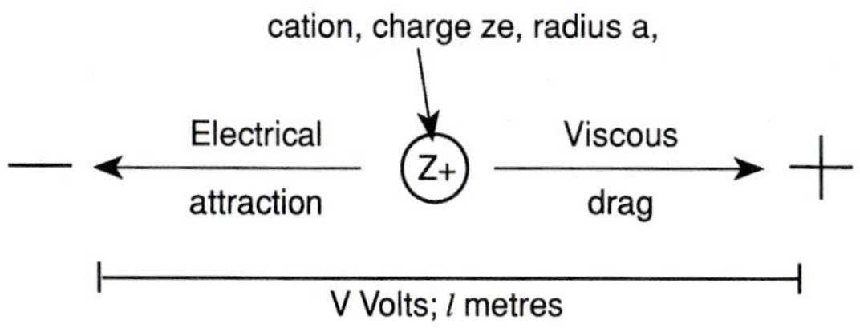

Fig. 3.7 The forces acting on a cation in an electric field.

As a rule of thumb 'macroscopic' spheres are probably larger than about 50 Å. The sphere needs to be large in comparison with the size of the solvent molecules. Clearly Stokes' law will not apply *exactly* to ions in solvents such as given in Table 3.2 since the ions will be comparable in size with the solvent molecules. Nevertheless the argument in the main text develops the basic physical principles governing the magnitude of single ion conductivities.

Table 3.2 Solvent viscosities at 25°C

Solvent	$\eta/10^{-3}$kg m^{-1}s^{-1}
H_2O	0.89
CH_3OH	0.54
Tetrahydrofuran	0.46
CH_3CN	0.34
Glycerol	$\approx 10^4$

At the same time the ion will suffer a *viscous drag* which will tend to retard the motion towards the negatively charged electrode. The viscous drag can be thought of as a frictional force due to the movement of the ion past and over solvent molecules which serve to impede the progress of the ion. The magnitude of the retarding force is difficult to estimate precisely but a useful estimate may be made using *Stokes' Law* which has been experimentally established for macroscopic spheres. In these cases the force is,

$$\text{Viscous Force} = 6\pi a v \eta \tag{3.11}$$

where a is the radius of the sphere, v is its velocity and η is a property of the medium known as its viscosity. Table 3.2 lists representative viscosities.

It can be seen that the frictional retarding force increases as the ion moves faster. As a result the ion will eventually move in solution with acceleration until it reaches a steady velocity v. Then the forces of attraction and repulsion balance each other. At this point

$$ze\left(\frac{V}{l}\right) = 6\pi a v \eta \tag{3.12}$$

Now the molar conductivity of the ion,

$$\Lambda \propto \frac{v}{(V/l)} \propto \frac{ze(V/l)}{6\pi a \eta (V/l)} \propto \frac{ze}{6\pi a \eta} \tag{3.13}$$

This predicts that Λ should be large for
- small ions
- highly charged ions
- solvents of low viscosity

It is interesting to compare the predictions of eqn (3.13) with the experimental data (Table 3.1). In respect of charge the model is good in that it correctly anticipates that the molar conductivities of the divalent alkaline earth metal ions Mg^{2+}, Ca^{2+} and Ba^{2+} are greater than those of the monovalent cations of the alkali metals Li^+, Na^+, K^+. However the model predicts

$$\Lambda_{Li+} > \Lambda_{Na+} > \Lambda_{K+}$$

which is the opposite trend to that reported in Table 3.1. This result is, at first sight, paradoxical since the radii of the gas phase ions follow the sequence

$$r_{Li+} < r_{Na+} < r_{K+}$$

However, as described in the previous chapter, the ions in solution are strongly hydrated and the extent of this follows the charge density of the ion. Consequently the sizes of the *hydrated* ions are

$$r_{Li+} > r_{Na+} > r_{K+}$$

which shows that the conductivity trend with the alkali metal cations is consistent with the Stoke's law model given provided the ions move to

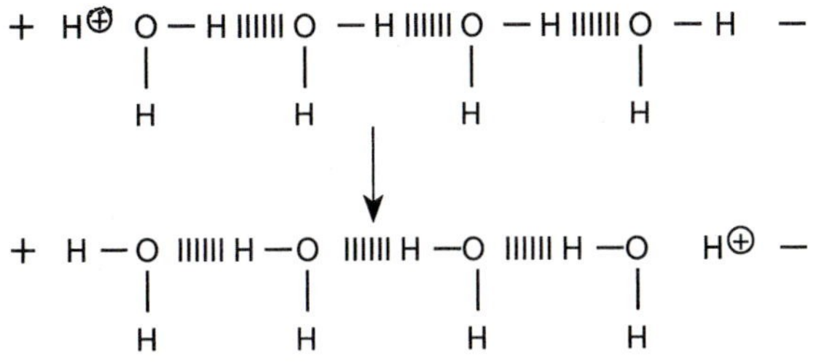

Fig. 3.8 Proton movement in water.

their respective electrodes with their hydration shells essentially intact, as is thought to be the case.

Finally, examination of Table 3.1 shows that H^+ and OH^- have anomalously large conductivities in aqueous solution, being substantially larger than found for simple singly charged ions such as K^+ or Br^-. This is because a specific conduction mechanism can operate for these two species in water. Whereas other ions are pulled to the electrode by coulombic attraction and move by pushing aside solvent molecules, the H^+ and OH^- ions can take advantage of the water that would otherwise inhibit their electrode-wards movements. The solvent is extensively hydrogen-bonded. As a result a proton encountering a chain of hydrogen-bonded solvent molecules can induce reorganisation of the hydrogen bonding as shown in Fig. 3.8 where ||||| denotes a hydrogen bond. The effect of this rearrangement is that the proton travels a distance of several water molecules without having to knock them out of its path. The proton effectively takes advantage of the hydrogen-bonded structure and exploits it to move towards the electrode with much less frictional resistance than is experienced by ions such as K^+ or Cl^-. Accordingly H^+ has an anomalously high molar conductivity as seen in Table 3.1. A similar mechanism operates for hydroxide ions in water (Fig. 3.9).

This special proton conduction mechanism is known as the Grotthuss mechanism.

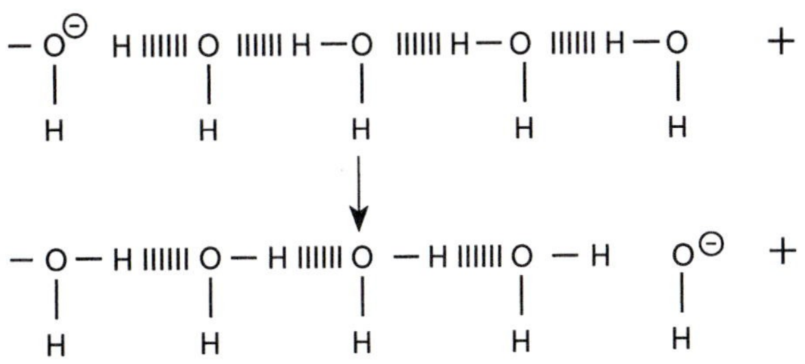

Fig. 3.9 Hydroxide ion movement in water.

3.6 Diffusion

In Section 3.1 we saw that ions in solution are induced to move by a voltage gradient. This gave rise to a current such that

$$\text{Current Density} = \kappa \times \text{Voltage Gradient} \tag{3.14}$$

where κ is the solution conductivity. The current was seen to be

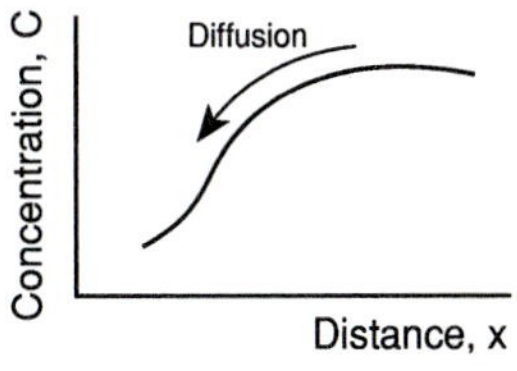

Fig. 3.10 Diffusion takes place down a concentration gradient.

composed of anions moving towards the anode and cations travelling independently in the opposite direction to the cathode.

Ions—and also uncharged molecules—may be induced to move in solution by means of an entirely different driving force to a voltage gradient, namely a *concentration gradient*. This is illustrated in Fig. 3.10 which shows a non-uniform concentration. There exists a concentration gradient. Intuition tells us that material will move ('diffuse') from high concentration to low concentration so as to 'even out' the concentration. In other words we would expect movement of material down the concentration gradient. In the case of Fig. 3.10 material would diffuse from right to left.

The rate of diffusion down a concentration gradient is quantified by Fick's first law of diffusion:

$$\text{Flux} = D \times \text{Concentration Gradient} \tag{3.15}$$

Mathematically the concentration gradient at any point in Fig. (3.10) is given by $\partial c / \partial x$ so that Fick's first law is flux $= D\partial c/\partial x$ (mol cm^{-2} s^{-1}).

The flux is the number of moles of material passing through unit area in a unit time and D is the diffusion coefficient of the moving species. Comparison of eqns (3.14) and (3.15) show that the current density and the flux effectively measure the *rate* of electrical conduction or of diffusion resulting from a *driving force*, of either a voltage gradient or a concentration gradient.

Regardless of whether an ion is induced to move by a concentration gradient or a voltage gradient the factors impeding the motion will be the same. Thus an ion which can conduct (migrate) rapidly will also diffuse rapidly. For example H^+ ions can use the special hopping (Grotthuss) mechanism described in the previous section. Equally the frictional resistance to a Cl^- ion moving through a solution resulting from having to displace solvent molecules will be similar for both migrating and diffusion chloride ions. As a consequence we would expect that the process of conduction and diffusion are closely related and this is reflected in the Nernst–Einstein equation:

$$D_{\mathrm{i}} = \Lambda_i \frac{k_{\mathrm{B}} T}{z^2 e^2} \tag{3.16}$$

which shows that the diffusion coefficient D_{i} of ion i is directly proportional to its molar conductivity Λ_{i}. The other terms in eqn (3.16) are k_{B}, the Bolztmann Constant, T, the absolute temperature, z, the ionic charge and e, the charge on an electron.

4 Going further

Chapters 2 and 3 covered the topics of activity coefficients and conductivity. We now return the study of electrode potentials as suspended at the end of Chapter 1.

4.1 Measurement of standard electrode potentials

We start our continuation by asking how we might measure standard electrode potentials. Suppose, for example, we wanted to obtain the value of the standard electrode potential of the cell:

$$\mathrm{Pt} \mid \mathrm{H_2}\ (\mathrm{g})\ (P = 1) \mid \mathrm{H^+}\ (\mathrm{aq})\ (a = 1)\ \mathrm{Cl^-}\ (\mathrm{aq})\ (a = 1) \mid \mathrm{AgCl} \mid \mathrm{Ag}$$

This cell is known as a Harned cell

This is, of course, equivalent to investigating the standard electrode potential of the Ag/AgCl couple:

$$E^{\ominus}_{\mathrm{Ag/AgCl}} = \phi_{\mathrm{Ag}} - \phi_{\mathrm{Pt}}$$

where ϕ_{Ag} and ϕ_{Pt} refer to the electrical potentials of the silver and platinum electrodes respectively. The factors discussed in chapter one lead us to recognise that the potential must be measured under conditions where no current flows. Thus a digital voltmeter is used for the measurement since this effectively draws negligible current. If we consider the further experimental arrangements involved for the cell of interest, we see that it is necessary to make the hydrogen gas of one atmosphere pressure and the solution unit activity in hydrochloric acid. The former requirement is readily realised but the latter will be difficult unless the experimenter knows the activity coefficient of hydrochloric acid as a function of its concentration in which case the relationships

$$a_{\mathrm{H^+}} = \gamma_{\mathrm{H^+}}\,[\mathrm{H^+}] \qquad a_{\mathrm{Cl^-}} = \gamma_{\mathrm{Cl^-}}\,[\mathrm{Cl^-}]$$
$$\gamma^2_{\pm} = \gamma_{\mathrm{H^+}}\,\gamma_{\mathrm{Cl^-}}$$

can be used to predict the concentration necessary to give unit activity. In practice a concentration of 1.18 M hydrochloric acid is required at 25°C for the activity to be unity. This implies $\gamma_{\pm} = 0.85$ and illustrates the observation made in Chapter 2 that electrolyte solutions can deviate appreciably from ideality. However, in general it is not usually the case that the concentrations corresponding to unit activity are known and thus it is necessary to proceed as follows.

The potential of the cell is given by:-

$$E = E^{\ominus}_{\mathrm{Ag/AgCl}} - (RT/F)\ln\{a_{\mathrm{H^+}}a_{\mathrm{Cl^-}}\} \qquad (4.1)$$

where the activities are related to the concentrations as described above by means of the activity coefficients $\gamma_{\mathrm{H^+}}$ and $\gamma_{\mathrm{Cl^-}}$ or $\gamma_{\pm}$. Now if the

$|z_+z_-|$ means the modulus or the positive value of the product z_+z_-.

solution is suitably dilute ($< 10^{-2}$ M) the Debye–Hückel limiting law can be used to predict the ionic activity coefficients:

$$\log_{10}\gamma_+ = -A z_+^2 \sqrt{I} \qquad \log_{10}\gamma_- = -A z_-^2 \sqrt{I}$$

or

$$\log_{10}\gamma_\pm = -A\,|z_+z_-|\sqrt{I} \tag{4.2}$$

where z_+ and z_- are the charges on the cation and anion respectively and I, the ionic strength of the solution is given by

$$I = \tfrac{1}{2}\sum_i c_i z_i^2 \tag{4.3}$$

In the example of interest $z_+ = z_- = 1$ and A is equal to 0.509 (in water at 25°C). Inserting the above into eqn 4.1 gives:

$$E = E^{\ominus}_{Ag/AgCl} - (RT/F)\ln[HCl]^2 - (RT/F)\ln\gamma_\pm{}^2 \tag{4.4}$$

which can be arranged to give

$$E + (2RT/F)\ln[HCl] = E^{\ominus}_{Ag/AgCl} - (2RT/F)\ln\gamma_\pm \tag{4.5}$$

On relating the activity coefficient to the ionic strength we finally obtain

Recall that $2.303\log_{10}x = \ln x$ so that $2.34 = 2.303 \times 0.509 \times 2$.

$$E + (2RT/F)\ln[HCl] = E^{\ominus}_{Ag/AgCl} + (2.34RT/F)\sqrt{[HCl]} \tag{4.6}$$

and consequently if we measure the cell potential, E, as a function of the HCl concentration in the region where the Debye–Hückel limiting law applies ($[HCl] \leq 10^{-2}$ M) a plot of $E + (2RT/F)\ln[HCl]$ against $\sqrt{[HCl]}$ should give a straight line with an intercept equal to the standard electrode potential of the Ag/AgCl couple. This is illustrated schematically in Fig. 4.1.

At higher concentrations it will be found that data will deviate from the straight line expected. This is duxe to the break down of the Debye–Hückel limiting law.

It is interesting that the above experiments were conducted on solutions of hydrochloric acid in the concentration range 0–10^{-2} M so that the Deybe–Hückel theory could be used as a basis for dealing with the unknown activity coefficients and hence eqn (4.6) used to rationalise the dependence of the measured cell potential on [HCl]. This procedure is typically that used for the accurate determination of standard electrode potentials and is *probably* quite different from what might be expected at first sight to be the relevant approach. This might have been the approximation of unit activity by unit molarity. This would have suggested measurement of the potential of the following cell for determining $E^{\ominus}_{Ag/AgCl}$:

$$Pt\,|\,H_2\,(g)\,(P = 1)\,|\,HCl\,(aq)\,(1\text{ M})\,|\,AgCl\,|\,Ag$$

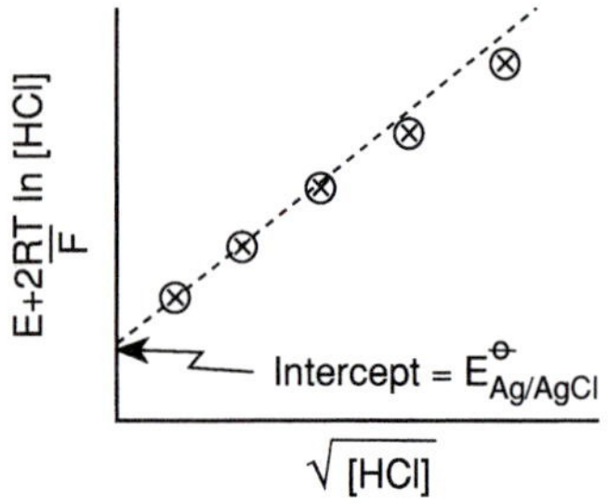

Fig. 4.1 Analysis of cell potential–concentration data from a Harned cell to permit the deduction of the standard electrode potential, $E^{\ominus}_{Ag/AgCl}$.

This is inappropriate because, as we have seen, a one molar solution of hydrochloric acid departs markedly from ideality.

The above experiment is a little special in the sense that both electrodes dip into the same electrolyte solution. The latter contains H^+ ions which participate (with the hydrogen gas) in establishing the potential at the

Box 4.1 The measurement of activity coefficients

In Chapter 2, the variation of activity coefficients with electrolyte concentration was discussed, Figure 2.8 showed the concentration dependence for several solutions. Following the discussion in Section 4.1 we can now see the basis for the experimental determination of solution non-ideality and the measurement of data such as that in Fig. 2.8. As an illustration consider the Harned cell,

$$\text{Pt} \mid H_2\,(g)\ \ (P = 1) \mid \text{HCl(aq)}\ (m) \mid \text{AgCl} \mid \text{Ag}$$

where m is the *molality* (moles kg^{-1}) of the dissolved hydrochloric acid. Experiment shows that the cell potential, E, depends on m in the following manner

m/mol kg^{-1}	0.1238	0.02563	0.009138	0.005619	0.003215
E/mV	341.99	418.24	468.60	492.57	520.63

The same cell has a potential of 352.4 mV when $m = 0.100$ mol kg^{-1}. Let us find the activity coefficient, $\gamma_\pm$, of 0.1 m hydrochloric acid, such that the activity,

$$a = \gamma_\pm m$$

The analysis of Section 4.1 leads us from the Nernst equation,

$$E = E^{\ominus}_{Ag/AgCl} - (RT/F)\ln a_{H^+} a_{Cl^-}$$

to the expression,

$$E + (2RT/F)\ln m = E^{\ominus}_{Ag/AgCl} + (2.34RT/F)\sqrt{m}$$

This expression is valid for molalities, m, such that the Debye–Hückel limiting law provides a quantitative prediction of the activity coefficients. We can use the data above to work out values of $[E + (2R/F)\ln m]$ and $\sqrt{m}$:

$(m/\text{mol kg}^{-1})^{1/2}$	0.352	0.160	0.0956	0.0750	0.0567
$[E + (2R/F)\ln m]/\text{m}V$	234.73	230.12	227.53	226.54	225.83

The resulting data are plotted in the figure below:

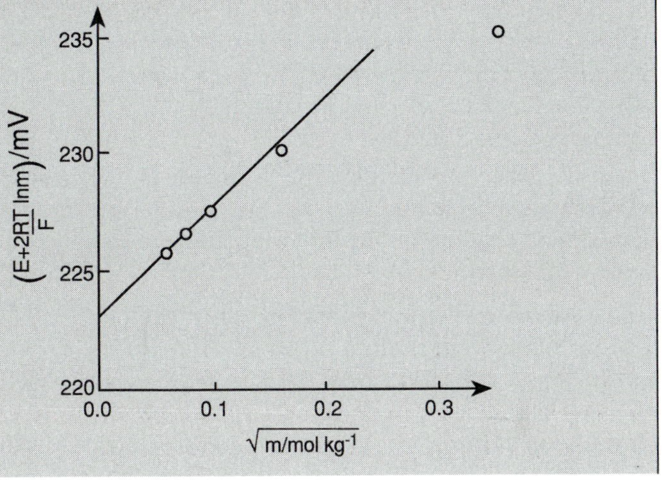

It can be seen that the low molality data gives the straight line predicted by the Debye–Hückel theory. Extrapolation of this line to infinitesimal molality ($m \rightarrow 0$) gives,

$$E^{\ominus}_{Ag/AgCl} = 223 \text{ mV}$$

We can now find $\gamma_{\pm}$ at ionic strengths where the limiting law is inapplicable. Consider the molality of 0.1 mol kg^{-1}. Substituting this value together with the corresponding observed cell potential (352.4 mV) and the now known value of $E^{\ominus}_{Ag/AgCl}$ into the Nernst equation,

$$E = E^{\ominus}_{Ag/AgCl} - (RT/F)\ln\{\gamma^2_{\pm}\, m^2\}$$

gives

$$0.3524 = 0.223 - (0.059)\log_{10}\{\gamma^2_{\pm}(0.1)^2\}$$

at 25°C, from which it may be deduced that

$$\gamma_{\pm} = 0.64$$

In general activity coefficients are found by making accurate measurements of cell EMFs and using Debye–Hückel type extrapolation of dilute solution data, as discussed in Section 4.1 to determine the corresponding standard cell potential. The latter combined with the Nernst equation, as above, allows the inference of $\gamma_{\pm}$ for concentrated solutions.

platinum electrode and also Cl^- ions which help produce the potential on the silver/silver chloride electrode. In other cases it is simply not possible to use a single solution. For example, suppose we wished to measure the standard electrode potential of the Fe^{2+}/Fe^{3+} couple. A single solution cell,

$$Pt \mid H_2\,(g)\,(P = 1) \mid H^+\,(aq)(a = 1),\ Fe^{2+}\,(aq)(a = 1),\ Fe^{3+}\,(aq)(a = 1) \mid Pt$$

is inappropriate since *both* platinum electrodes are exposed to the Fe^{2+} and Fe^{3+} ions. Thus at one platinum electrode the appropriate potential determining equilibrium is set up:

$$Fe^{3+}\,(aq) + e^-\,(metal) \rightleftharpoons Fe^{2+}\,(aq)$$

but at the other *both* the Fe^{2+}/Fe^{3+} and the H^+/H_2 couples will try and establish their potentials so that in addition to the above potential determining equilibrium the following different equilibrium is also trying to be established

$$H^+\,(aq) + e^-\,(metal) \rightleftharpoons \tfrac{1}{2}H_2\,(g)$$

In practice the fact that both redox couples are exchanging electrons readily with the electrode will mean that the downhill reaction ($\Delta G^{\ominus} = -0.77F$)

$$Fe^{3+}\ (aq) + \tfrac{1}{2}H_2\ (g) \rightleftharpoons Fe^{2+}\ (aq) + H^+\ (aq)$$

will be catalysed by the electrode and so the Fe^{3+} ions in solution will be reduced by the hydrogen gas.

It follows that if sensible measurements are to be made to find the standard electrode potential of the Fe^{2+}/Fe^{3+} couple, the two '*half cells*' must be separated using a salt bridge as shown in Fig. 4.2 and summarised by the cell notation

$$Pt \mid H_2\ (g)\ (P=1) \mid H^+\ (aq)(a=1) \parallel Fe^{2+}\ (aq)(a=1),\ Fe^{3+}\ (aq)(a=1) \mid Pt$$

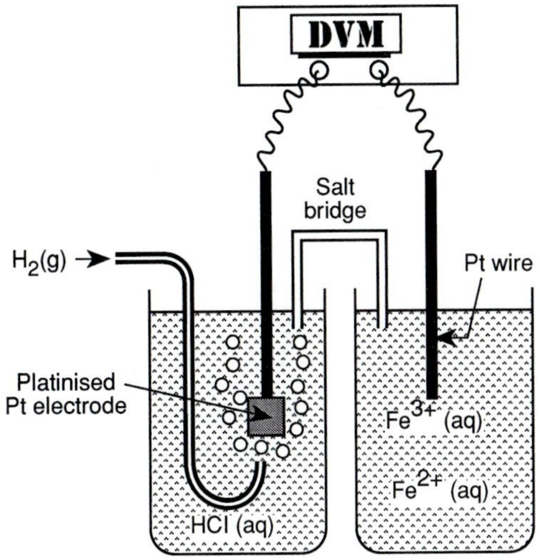

Fig. 4.2 A salt bridge separates the two half cells to ensure no mixing of the respective solutions.

4.2 Salt bridges

We have seen in the previous section that in most cases it is not possible to put two half cells directly into contact when we wish to make a measurement of a cell potential. Therefore, to avoid this, the two half cells are set up in separate containers and connected by a *salt bridge* containing an aqueous solution of potassium chloride as illustrated in Fig. 4.3.

Rather than using a solution of potassium chloride in the salt bridge it is often more convenient (less messy!) to employ a gel containing the same electrolyte.

Figure 4.3 shows the half cells for the Cu/Cu^{2+} and the Zn/Zn^{2+} couples separated by a salt bridge containing saturated (approximately 5 M) aqueous potassium chloride solution. If the couples were not

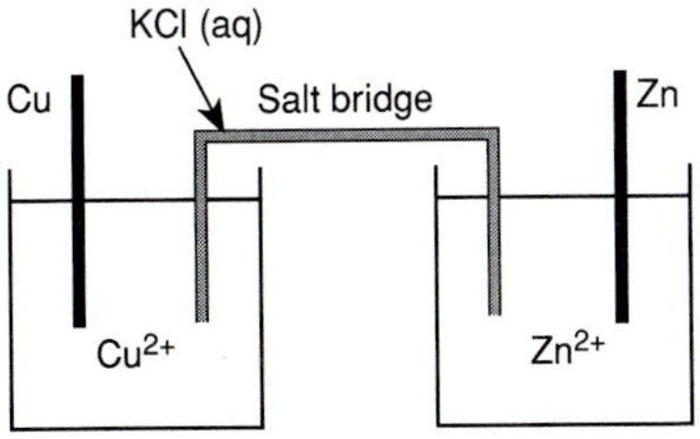

Fig. 4.3 A salt bridge links the Cu/Cu^{2+} and Zn/Zn^{2+} half cells.

separated but held in a single container then the following reaction

$$Zn\,(s) + Cu^{2+}\,(aq) \rightleftharpoons Zn^{2+}\,(aq) + Cu\,(s) \qquad \Delta G^{\circ} = -1.11F$$

would occur and the zinc electrode would very rapidly become copper plated, rather destroying the original idea of the experiment.

It will be noted that we have chosen to use an aqueous solution of KCl as the electrolyte in the salt bridge. A solution of ammonium nitrate would have been equally satisfactory. However the vast majority of other electrolytes would not. To understand which salts are appropriate, and which are not, requires a brief appreciation of *liquid junction potentials*, to which we next turn.

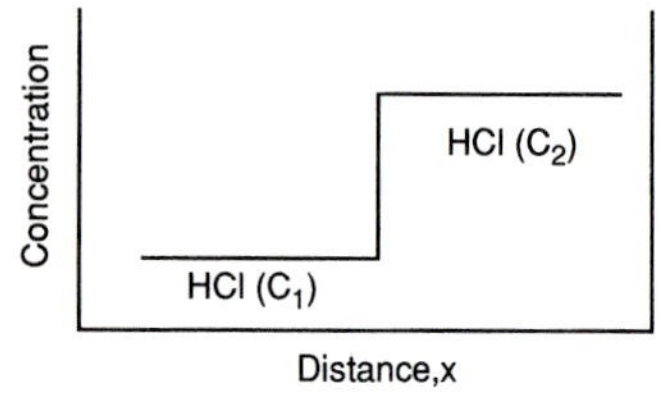

Fig. 4.4

The origin of these may be understood by considering two HCl solutions of different concentrations c_1 and c_2 put in contact, initially as shown schematically in Fig. 4.4. Clearly there will be a large concentration gradient, $\partial[\mathrm{HCl}]/\partial x$, at the interface between the two solutions. Such a concentration gradient will lead to *diffusion* of both H^+ and Cl^-, as outlined in Chapter 3, from high concentration to low concentration so as to tend to equalise the two concentrations.

In Chapter 3 it was seen that the rate of diffusion is measured by the flux of the diffusing species. That is the number of moles passing though unit area in unit time. Thus if we focus on the interface between the two solutions depicted in Fig. 4.4 we can write for the flux, j_{H^+}, of the protons

$$j_{H^+} = D_{H^+}\frac{\partial[H^+]}{\partial x} \qquad (4.7)$$

and for the flux of the chloride ions,

$$j_{Cl^-} = D_{Cl^-}\frac{\partial[Cl^-]}{\partial x} \qquad (4.8)$$

Initially shortly after the interface is formed,

$$\frac{\partial[H^+]}{\partial x} = \frac{\partial[Cl^-]}{\partial x} \qquad (4.9)$$

But we know from Chapter 3 that D_{H^+} is much larger than D_{Cl^-} since the proton can diffuse using the facile Grotthuss mechanism whereas the chloride ion has to displace solvent molecules and so diffuses more slowly. It follows that initially the protons will diffuse at a faster rate than the chloride ions. As a result of this a charge difference and hence a potential difference will be set up across the interface between the two solutions. The solution of lower concentration will become positively charged because of the 'gain' of protons whereas the higher concentration will become negatively charged. This will have the effect that the rate of the chloride ion transport will be accelerated (since migration will now contribute to the rate of transport) and the proton transport rate will be retarded. Ultimately a *steady state* will be reached as shown Fig. 4.5. Nevertheless at this steady state a potential difference will exist at the boundary of the two solutions (Fig. 4.5). This is known as a *liquid*

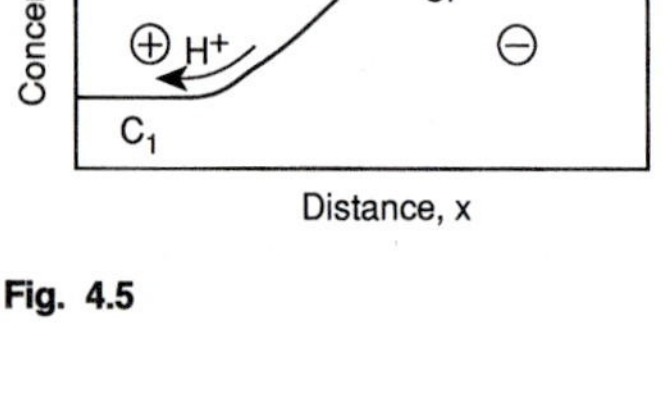

Fig. 4.5

junction potential. The size of this is given by the equation

$$E_{\text{LJP}} = (t_+ - t_-)\frac{RT}{F}\ln\left(\frac{c_2}{c_1}\right) \quad (4.10)$$

A full derivation of eqn (4.10) is given, starting on page 16 of *Electrochemistry*, P. Rieger, Prentice-Hall, 1987.

where t_+ and t_- are transport numbers of the two ions. In the case illustrated above t_+ refers to H^+ and t_- to Cl^-.

The form of eqn (4.10) can be rationalised as follows. First, the transport numbers are proportional to the ionic conductivity and hence the diffusion coefficients of the two ions as explained in Chapter 3. Thus the larger the difference in the transport numbers of the two ions the greater will be the liquid junction potential expected. Secondly, the liquid junction potential depends on the ratio c_2/c_1; if the two concentrations are equal the potential is predicted to disappear as would be anticipated.

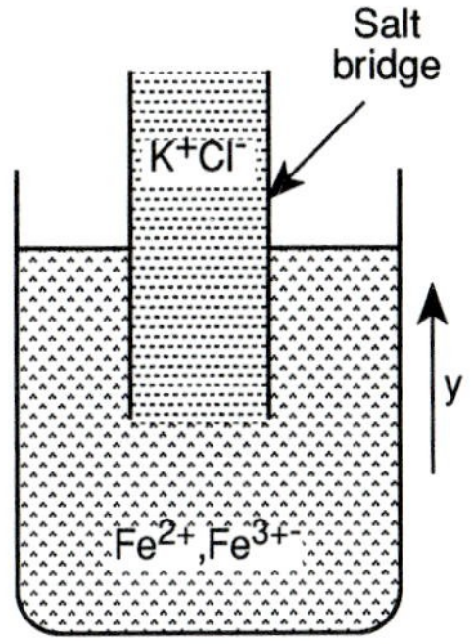

Fig. 4.6

Equation (4.10) predicts the magnitude of the liquid junction potential between two solutions when conditions pertain of steady state transport between them. Let us now return to the question of which electrolytes are appropriate to use in a salt bridge. Figure 4.6 shows one end of a salt bridge dipping into a solution of Fe^{2+} and Fe^{3+} as in Fig. 4.2. The salt bridge will contain saturated aqueous KCl which corresponds approximately to a concentration of 5 M. The half cell will contain Fe^{2+} and Fe^{3+} at concentrations below about 10^{-2} M if it is being used to measure standard electrode potentials (see previous section). Consequently there will be a large amount of diffusion of K^+ and Cl^- out of the salt bridge but a tiny amount of diffusion of Fe^{2+} and Fe^{3+} in the opposite direction.

Recall that the fluxes of diffusing species are proportional to their concentration gradients and since at the interface

$$\frac{\partial[K^+]}{\partial y} \sim \frac{\partial[Cl^-]}{\partial y} >> \frac{\partial[Fe^{3+}]}{\partial y} \sim \frac{\partial[Fe^{2+}]}{\partial y}$$

the fluxes of K^+ and Cl^- will be vastly greater than those of Fe^{2+} or Fe^{3+}.

As a result of the ions K^+ and Cl^- diffusing out of the salt bridge a liquid junction potential can be set up. The size of this will be proportional to the difference of their transport numbers

$$E_{\text{LJP}} \propto (t_{K^+} - t_{Cl^-})$$

However Table 3.1 shows that the two ions conduct and hence diffuse at almost exactly the same rates. As a result

$$t_{K^+} \sim t_{Cl^-} \sim 0.5$$

so that the liquid junction potential vanishes.

It can now be appreciated why KCl was selected as an electrolyte of choice for a salt bridge: *it gives rise to negligible liquid-junction potentials.* Imagine, in contrast, what would happen if the bridge contained ions other than K^+ or Cl^-, say H^+ and Cl^-. This is depicted in Fig. 4.7. In this case there would now be significant fluxes of H^+ and Cl^- diffusing out of the salt bridge, but since

$$t_{H^+} >> t_{Cl^-}$$

an appreciable liquid junction potential would develop. If we were attempting to measure the standard electrode potential of the Fe^{2+}/Fe^{3+} couple using the cell

$$\text{Pt(2)} \mid H_2\,(g)\,(P = 1) \mid \text{HCl (aq)}(a = 1) \parallel Fe^{2+}\,(aq)(a = 1),\; Fe^{3+}\,(aq)(a = 1) \mid \text{Pt(1)}$$

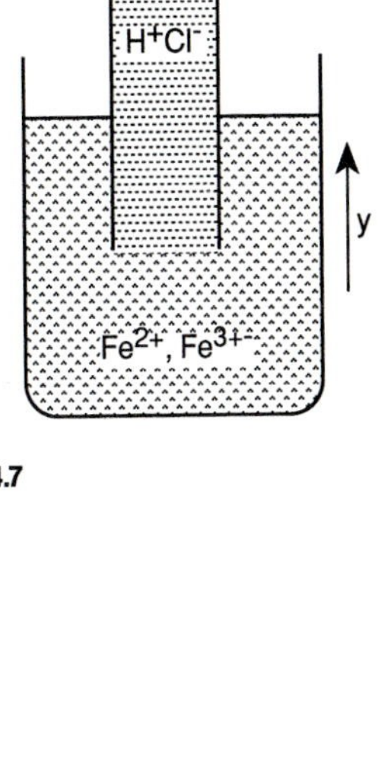

Fig. 4.7

then, if the salt bridge contained HCl, we would measure a potential difference

$$\phi_{Pt(1)} - \phi_{Pt(2)} = E^{\ominus}_{Fe^{2+}/Fe^{3+}} + E_{LJP}$$

where E_{LJP} represents the potential at the liquid–liquid junction shown in Fig. 4.7. In contrast when the salt bridge contains KCl there will be no liquid junction potential and a true measurement of $E^{\ominus}$ will result.

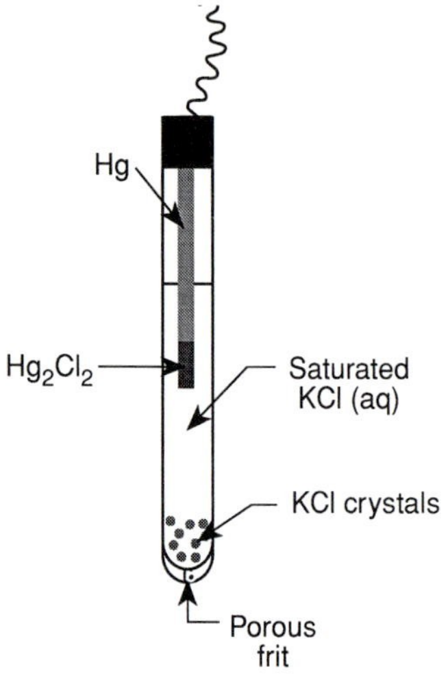

Fig. 4.8 A saturated calomel reference electrode.

4.3 The calomel reference electrode

In Chapter 1 we introduced the standard hydrogen electrode as the reference electrode against which all other half cells are measured and reported. This does not, however, dictate that the hydrogen electrode has to be used *experimentally*. Indeed it is often inconvenient to use since the hydrogen gas has to be safely vented, the platinum electrodes freshly plated with platinum black before use and the solution of hydrochloric acid employed very accurately made up to a concentration corresponding to unit activity. Instead it is much more convenient to use a *calomel electrode* as the reference electrode. This electrode is shown in Fig. 4.8. The potential determining equilibrium is

$$\tfrac{1}{2}Hg_2Cl_2(s) + e^- \rightleftharpoons Hg(l) + Cl^-(aq)$$

so that the potential developed between the mercury metal (M) and the solution (S) is

$$\phi_M - \phi_S = \text{const} - \frac{RT}{F}\ln a_{Cl^-}$$

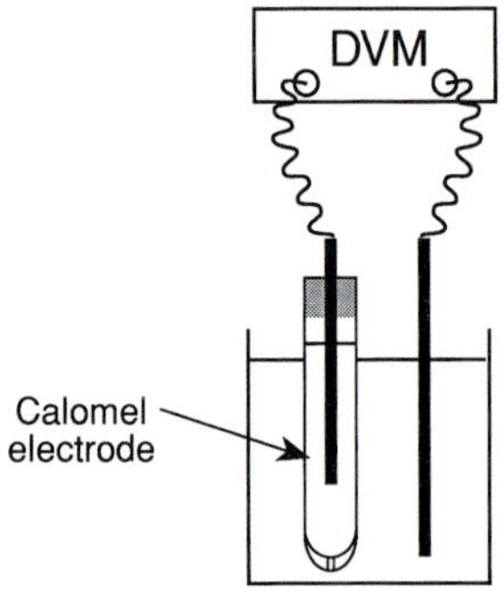

Fig. 4.9 A calomel electrode used as a reference electrode eliminates liquid junction potentials.

It can be seen that the potential of the calomel electrode is dependent on the chloride ion activity. This, however, is not inconvenient as the potassium chloride solution inside the electrode is maintained at a saturated, and therefore constant level by the KCl crystals present. The electrode contacts the solution by means of a porous frit which in effect acts as a salt bridge. This means that when measurements are made, as shown in Fig. 4.9, they are essentially free of any liquid junction problems automatically.

The electrode potential of the Hg_2Cl_2/Hg couple in contact with a saturated solution of chloride ions is 0.242 V at 25°C. Accordingly if measurements are made relative to a saturated calomel reference electrode—as in Fig. 4.9—they are readily correlated to standard hydrogen reference electrode scale by adding this value to the number measured

$$E^{\ominus}_{H_2/H^+} = E^{\ominus}_{calomel} + 0.242$$

Notice that the value of 0.242 is not the *standard* electrode potential of the Hg/Hg_2Cl_2 couple since the activity of Cl^- in saturated aqueous KCl is not unity. In fact the standard potential is +0.27 V whilst if the calomel electrode contains 1 M KCl the potential is +0.283

4.4 Measurement of the standard electrode potentials of reactive metals

We have seen in Chapter 1 and above, that the value of the standard electrode potential of M/M^{n+} couples may be found from the cell

$$Pt \mid H_2\,(g)\,(P = 1) \mid H^+\,(aq)\,(a = 1) \parallel M^{n+}\,(aq)\,(a = 1) \mid M$$

Alternatively if a knowledge of the activity coefficient of M^{n+} (aq) is not available, measurements of the cell potential for

$$\text{Pt} \mid \text{H}_2\,(\text{g})\,(P = 1) \mid \text{H}^+\,(\text{aq})\,(a = 1) \parallel \text{M}^{n+}\,(\text{aq})\,(c) \mid \text{M}$$

are made as a function of the concentration of M^{n+}, c, in the range 0–10^{-2} M so that a Debye–Hückel extrapolation may be carried out as in Section 4.1 to find the sought standard electrode potential.

However this approach is not invariably feasible. For example, if we wish to evaluate the standard electrode potential of the Na^+/Na couple it is clear that a sodium electrode would have such a short lifetime in contact with an aqueous solution as to preclude any chance of making appropriate measurements. Instead we have to rely on an indirect, two step approach to the problem.

In step 1 the following cell is studied

$$\text{Na(s)} \mid \text{NaI}\,(\text{EtNH}_2) \mid \text{Na (amalgam, 0.606\%)}$$

in which the sodium iodide electrolyte is dissolved in the non-aqueous solvent ethylamine which is rigorously purified of trace water. One electrode is made from pure sodium and the other from a dilute solution of sodium in mercury (an '*amalgam*'). The potential determining equilibria are

$$\text{Na}^+\,(\text{EtNH}_2) + e^- \rightleftharpoons \text{Na (amalgam)}$$

and

$$\text{Na}^+\,(\text{EtNH}_2) + e^- \rightleftharpoons \text{Na (s)}$$

The formal cell reaction is thus

$$\text{Na (s)} \rightarrow \text{Na (amalgam)}$$

and the cell potential, E_1, is related to the free energy of transfer of sodium from its standard state (as sodium metal) to the amalgam.

In step 2 another cell is examined:

$$\text{Pt} \mid \text{H}_2\,(P = 1) \mid \text{HCl (aq)}(a = 1) \parallel \text{NaOH (aq)(0.2 M)} \mid \text{Na (amalgam 0.606\%)}$$

This contains a standard hydrogen electrode connected via a salt bridge with a half cell containing the same dilute amalgam as used in step 1, but now in contact with an aqueous solution of sodium hydroxide. If sodium metal were placed in contact with the latter it would be highly reactive. However the amalgam displays a kinetic stability and potential measurements are readily conducted on the electrode. In step two the potential determining equilibria at the two electrodes are

$$\text{Na}^+\,(\text{aq}) + e^- \rightleftharpoons \text{Na (amalgam)}$$

and

$$\text{H}^+\,(\text{aq}) + e^- \rightleftharpoons \tfrac{1}{2}\text{H}_2\,(\text{g})$$

The stability of the amalgam/water interface is strictly *kinetic*. In *thermodynamic* terms it would be favourable for the sodium dissolved in the mercury to react:

$$\text{Na (Hg)} + \text{H}_2\text{O} \rightarrow \text{Na}^+\,(\text{aq}) + \text{OH}^-\,(\text{aq}) + \tfrac{1}{2}\text{H}_2\,(\text{g})$$

This corresponds to a formal cell reaction of

$$Na^+ (aq) + \tfrac{1}{2}H_2 (g) \rightarrow Na (amalgam) + H^+$$

Suppose that the measured cell potential is E_2. Consider next the physical significance of the voltage $E_2 - E_1$. This must be related to the chemical reaction

$$Na^+ (aq) + \tfrac{1}{2}H_2 (g) \rightarrow Na (amalgam) + H^+(aq)$$

minus $$Na (s) \rightarrow Na (amalgam)$$

That is, to

$$Na^+ (aq) + \tfrac{1}{2}H_2 (g) \rightarrow Na (s) + H^+(aq)$$

which is the formal cell reaction of the hypothetical cell

$$Pt \mid H_2 (P = 1) \mid H^+ (aq)(a = 1) \parallel Na^+ (aq) \mid Na$$

So by the Nernst equation

$$(E_2 - E_1) = E^{\ominus}_{Na/Na^+} + \tfrac{RT}{F} \ln a_{Na^+}$$

where, $a_{Na^+} = \gamma_{Na^+} [Na^+] = 0.2\ \gamma_{Na^+}$. It is known that that $\gamma_{Na^+} = 0.756$ in 0.2 M NaOH so measurements of E_1 and E_2 give a value for $E^{\ominus}_{Na/Na^+}$. As might be expected this is very negative,

The measurement of activity coefficients is described in Box 4.1.

$$E^{\ominus} = -2.712 \text{ V}$$

5 Further applications

This chapter briefly highlights some applications of electrode potentials in addition to those that have emerged previously. We start by considering *membrane potentials*.

5.1 Membrane potentials

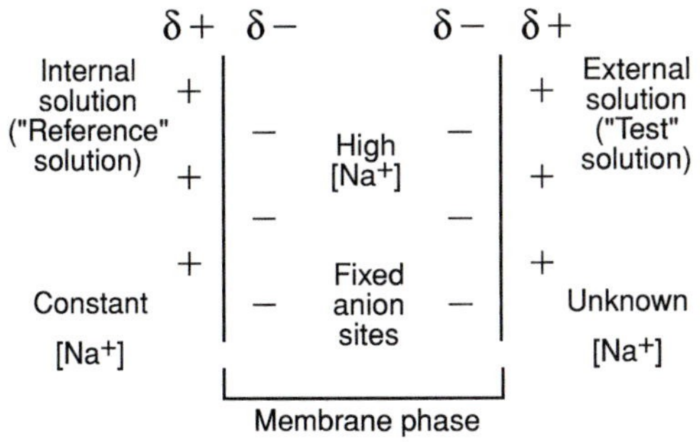

Fig. 5.1 A membrane which is permeable to sodium ions. It contains fixed anion sites.

Figure 5.1 shows an illustration of a cation exchange membrane. In this example only sodium cations can pass through the membrane; it is impermeable to the solvent, to anions and to all cations except Na^+. The membrane may contain fixed (immobile) anionic sites with which cations in the membrane are to some extent loosely chemically bonded.

If both sides of the membrane are exposed to (different) solutions containing sodium cations then we may expect the following equilibrium to be established at *each* solution/membrane interface:

$$Na^+(aq) \rightleftharpoons Na^+(membrane)$$

The chemical bonding of the cation may take the form of electrostatic attraction, as with protons in the glass electrode, or complexation as in the K^+ sensing microelectrode based on valinomycin. Both these type of membranes are discussed later in this chapter.

Of course both aqueous solutions will, for reasons of electroneutrality, contain anions but these are assumed to be excluded from the membrane. If this is the case, then the establishment of the above equilibrium will lead to a charge separation across each of the interfaces, in particular if the equilibrium lies to the left, sodium cations will tend to leave the membrane which will become negatively charged relative to the solution phase so that the charge separation will be as shown schematically in Fig. 5.1. Conversely, if the equilibrium lies to the right the membrane will become positively charged relative to the solution.

If the solutions (a 'reference' solution of fixed Na^+ concentration and a 'test' solution of variable Na^+ concentration) bathing each side of the membrane contain equal concentrations of sodium cations then the magnitude of the charge separation at each interface will be equal. However if the concentrations are unequal this will not be the case and a potential will be developed across the membrane. At equilibrium the electrochemical potentials of the Na^+ cation in the two solutions must be equal so that,

$$\bar{\mu}_{Na^+,Test} = \bar{\mu}_{Na^+,Reference} \tag{5.1}$$

But,

$$\bar{\mu}_{Na^+,Test} = \mu_{Na^+,Test} + F\phi_{Test} \tag{5.2}$$

and

$$\bar{\mu}_{Na^+,Reference} = \mu_{Na^+,Reference} + F\phi_{Reference} \tag{5.3}$$

where

$$\mu_{Na^+} = \mu^{\ominus}_{Na^+} + RT\ln[Na^+] \tag{5.4}$$

It follows that the membrane potential is given by

$$\begin{aligned}\Delta\phi_{Membrane} &= \phi_{Test} - \phi_{Reference} \\ &= -\left(\frac{RT}{F}\right)\ln\left(\frac{[Na^+]_{Test}}{[Na^+]_{Reference}}\right)\end{aligned} \tag{5.5}$$

Such potentials can be measured and exploited by locating reference electrodes such as calomel electrodes in each of the solution phases, using the following arrangement:

Internal reference electrode	Internal Solution (constant composition)	Membrane	External Solution	External reference electrode

In this way the potential difference measured between the two reference electrodes (R.E.s) is

$$\phi_{External\ R.E.} - \phi_{Internal\ R.E.} = A - \left(\frac{RT}{F}\right)\ln\left(\frac{[Na^+]_{External}}{[Na^+]_{Internal}}\right) \tag{5.6}$$

where the constant, A, is the potential difference (if any) observed when both the internal ('reference') and external ('test') solutions contain the same concentrations of sodium cations. If the internal solution is maintained at a fixed composition then the measured potential is,

$$\phi_{External\ R.E.} - \phi_{Internal\ R.E.} = B - \left(\frac{RT}{F}\right)\ln[Na^+]_{External} \tag{5.7}$$

where

$$B = A + \left(\frac{RT}{F}\right)\ln[Na^+]_{Internal} \tag{5.8}$$

Equation (5.7) shows that membrane potentials detected via a pair of reference electrodes, either side of an *ion–selective* membrane, can be used for analytical purposes to detect unknown concentrations of certain ions. The next two sections illustrate the concept in practice with reference to the detection of K^+ and H^+.

5.2 Potassium-selective microelectrodes

Potassium *micro*electrodes are suitable for the determination of K^+ ions in living tissues and in the intracellular space of individual cells.

Figure 5.2 shows a K^+ sensing electrode designed around a micropipette so that it can interrogate potassium concentrations in small areas, for example in living tissues. The tip of the micropipette contains a liquid membrane. The terminal 200 μm or so of the pipette is coated with silicone to keep this in place. The membrane is composed of a solution of the cyclic peptide *valinomycin* dissolved in diphenylether. As shown in Fig. 5.3 the peptide contains a cavity

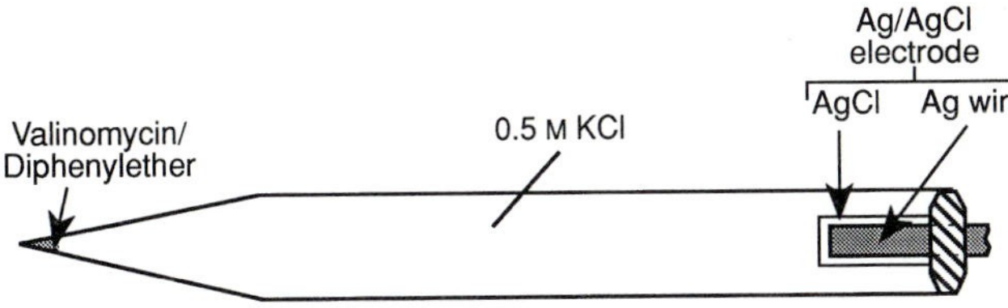

Fig. 5.2 A potassium selective microelectrode.

in which K^+ cations can be strongly completed. Complexation with K^+ is encouraged both by the precise size of the cavity and by the presence of carbonyl oxygens in the valinomycin ring which can participate in K^+ coordination without great conformational changes. As a result the peptide shows a high *selectivity* toward binding K^+ rather than any other cation.

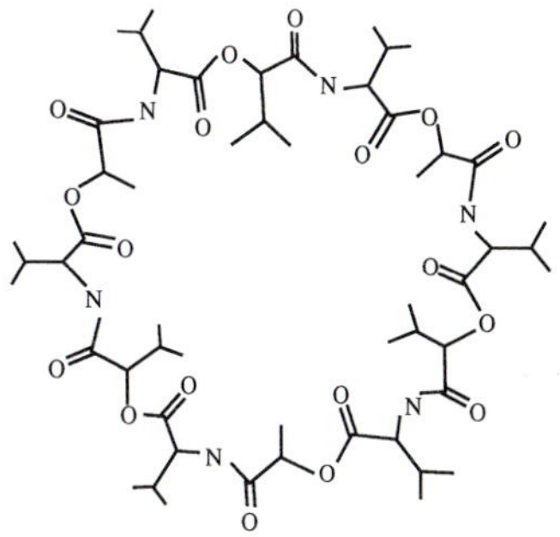

Fig. 5.3 The structure of valinomycin.

Inspection of Fig. 5.2 shows that inside the micropipette is an aqueous solution of potassium chloride which bathes a silver/silver chloride reference electrode. Accordingly when the micropipette is placed in an external solution containing an unknown level of K^+ ions and a second reference electrode is located in this external solution, a membrane potential is developed across the liquid (diphenylether) membrane. The potential difference measured between the internal silver/silver chloride electrode and the external reference electrode will be,

$$\phi_{\text{External R.E.}} - \phi_{\text{Internal R.E.}} = A' - \frac{RT}{F}\ln[K^+]_{\text{External}} \qquad (5.9)$$

This provides the basis for the *selective potentiometric* determination of unknown concentrations of K^+ ions. Potassium valinomycin electrodes and related sensors are commercially available and have been used to determine potassium ion concentrations in soils, sea-water, blood plasma and kidney tubules.

The complexation of metal ions, M, by valinomycin, Val, can be characterised by the equilibrium,

$$M + Val \rightleftharpoons M{\cdot}Val$$

and the stability constant

$$K_{val}(M) = \frac{[M{\cdot}Val]}{[M]{\cdot}[Val]}$$

For potassium cations in methanol solutions,

$$\log_{10}K_{val}(K^+) = 4.5$$

whereas for sodium cations,

$$\log_{10}K_{val}(Na^+) = 0.7$$

or for ammonium ions,

$$\log_{10}K_{val}(NH_4^+) = 1.7$$

The complexation with K^+ is therefore 3–4 orders of magnitude stronger than with Na^+ or NH_4^+.

5.3 Proton-selective glass electrodes

Measurements of the pH of aqueous solutions are vital in a broad diversity of area ranging from environmental monitoring to clinical chemistry. pH is a measurement of the *acidity* of a solution defined by

$$pH = -\log_{10}a_{H_3O^+} \approx -\log_{10}[H_3O^+] \qquad (5.10)$$

where $a_{H_3O^+}$ is the activity and $[H_3O^+]$ the concentration of free protons in the solution. Figure 5.4 shows a glass electrode which can be used to measure

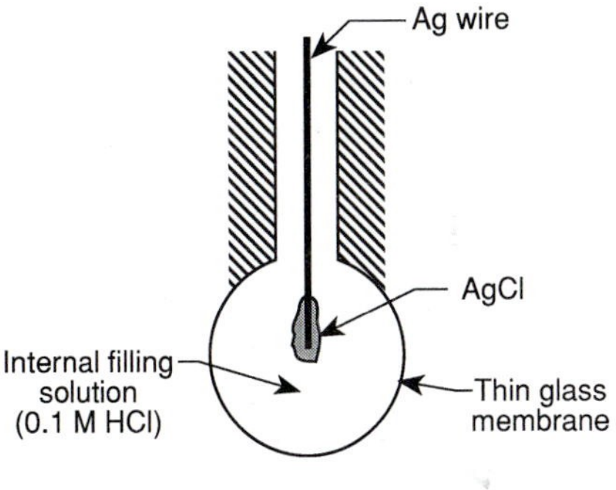

Fig. 5.4 A glass membrane electrode.

Many commercial pH electrodes comprise a combination of a glass membrane electrode (with its internal reference electrode) and an external reference electrode in a single probe.

pH over a range of 1–9. It comprises a thin (and fragile) glass membrane which is filled with an aqueous solution of hydrochloric acid. A silver/silver chloride electrode dips into the HCl and acts as an internal reference electrode. When dipped into an external solution, of unknown pH, containing a suitable reference electrode, the potential difference recorded between the two reference electrodes reflects a potential developed across the thin glass membrane. This arises since the glass functions as an ion-exchange resin and an equilibrium is established between Na^+ cations in the *surface* of the glass matrix and hydrogen ions in solution,

$$H_3O^+(aq) + Na^+(surface) \rightleftharpoons H_3O^+(surface) + Na^+(aq)$$

Note that the exchange is restricted to near the surface of the glass membrane where it is in contact with the internal or test solutions and where there exists a 'hydrated zone' of the glass as shown in Fig. 5.5.

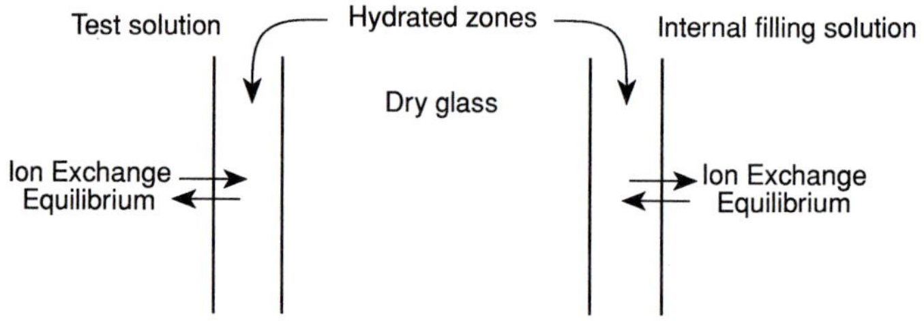

Fig. 5.5 Ion exchange equilibria are restricted to hydrated layers on the surface of glass membrane electrodes.

The potential difference between the internal and external reference electrodes is given by

$$\begin{aligned}\phi_{\text{External R.E.}} - \phi_{\text{Internal R.E.}} &= A'' - \frac{RT}{F}\ln[H^+]_{\text{External}} \\ &= A'' - \frac{2.303RT}{F}\log_{10}[H^+]_{\text{External}} \qquad (5.11) \\ &= A'' + \frac{2.303RT}{F}\text{pH}\end{aligned}$$

It follows that the recorded potential difference changes by 59 mV (at 25°C) for every unit change in pH.

The nature of the equilibrium which establishes the glass membrane potential suggest that high concentrations of Na^+ cations in a test solution might *interfere* with pH measurements since they will inhibit the H_3O^+/Na^+ exchange reaction. Such *alkali errors* are well known and lead to unreliable pH readings in test solutions under alkali conditions, pH > 9, in the presence of significant amounts of sodium ions. However the use of special glasses, for example lithium glass minimises this problem and can extend the reliable range of pH measurements to beyond pH 12.

The ability to sensitively and precisely detect acidity over the range from pH 1 to beyond pH 12 is exploited in other potentiometric analytical devices, namely *gas sensing electrodes,* and *enzyme electrodes* which can detect a wide range of substrates as is described in the next two sections.

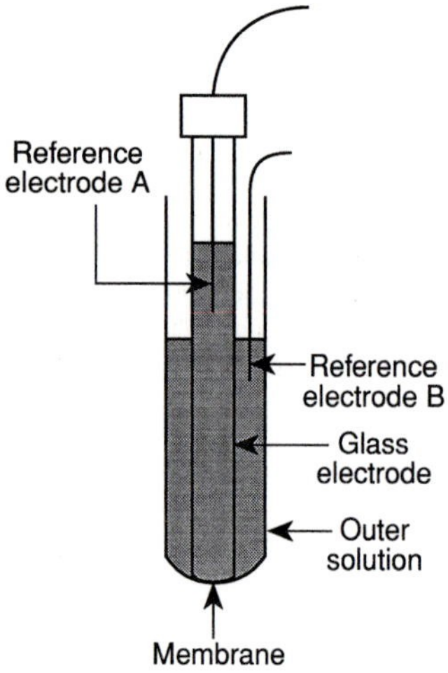

Fig. 5.6 A gas sensing electrode.

5.4 Carbon dioxide and ammonia gas sensing electrodes

A gas sensing probe based on a glass electrode is illustrated in Fig. 5.6. The pH sensitive glass electrode (with its own internal reference electrode, A) is itself contained within a further tube which terminates in a gas-permeable membrane. The space between the glass electrode and the outer tube is filled with an appropriate solution and contains a second reference electrode, B. If the outer solution is chosen so that its pH is influenced by the gas entering the device through the membrane, then measurements of the glass electrode potential relative to the outer reference electrode will provide an indirect measurement of the gas concentration. For example, if it is utilised to detect carbon dioxide, CO_2, the outer solution is selected to be aqueous 10^{-2} M $NaHCO_3$. The membrane is chosen to pass gaseous CO_2 from the ambient atmosphere so that relatively rapidly an equilibrium composition of dissolved CO_2 is established in the outer solution.

$$CO_2(g) \rightleftharpoons CO_2(aq)$$

The following equilibria are then also set up in the outer solution:

$$CO_2(aq) + H_2O(l) \rightleftharpoons H_2CO_3(aq)$$

$$H_2O(l) + H_2CO_3(aq) \rightleftharpoons H_3O^+(aq) + HCO_3^-(aq)$$

It follows therefore that the partial pressure of carbon dioxide, CO_2, in the gas is related to the H_3O^+ and HCO_3^- concentrations in the solutions by the equilibrium constant,

$$K = \frac{[H_3O^+][HCO_3^-]}{P_{CO_2}}$$

Since HCO_3^- is maintained at a much higher concentration ($\sim 10^{-2}$ M) than the levels of dissolved gas so that $[HCO_3^-]$ is nearly constant, it follows that

$$P_{CO_2} \propto [H_3O^+]$$

Accordingly the potential measured between the two reference electrodes (B and A in Fig. 5.6) is given by

$$\begin{aligned} \phi_B - \phi_A &= \text{constant} + \frac{2.303RT}{F}\text{pH} \\ &= \text{another constant} - \frac{2.303RT}{F}\log_{10}P_{CO_2} \end{aligned} \quad (5.12)$$

The gas probe therefore responds in a Nernstian manner to the partial pressure of CO_2. Levels as low as 10^{-5} M CO_2 can be detected in this way.

If the 10^{-2} M $NaHCO_3$ solution is replaced by 10^{-1} M NH_4Cl then an ammonia (NH_3) gas sensing electrode results. The key equilibria in this case are

A higher concentration of NH_4Cl is required as compared to $NaHCO_3$, since NH_3 is more soluble than CO_2 so higher levels of NH_4^+ than HCO_3^- are needed to keep these ions at effectively fixed concentrations.

$$NH_3(g) \rightleftharpoons NH_3(aq)$$

$$NH_3(aq) + H_3O^+(aq) \rightleftharpoons NH_4^+(aq) + H_2O(l)$$

and the resulting potential difference is

$$\phi_B - \phi_A = \text{constant} - \frac{2.303RT}{F}\log_{10}P_{NH_3} \quad (5.13)$$

where P_{NH3} is the ammonia partial pressure. Concentrations as low as 10^{-6} *M* of NH_3 may be measured.

5.5 Enzyme electrodes

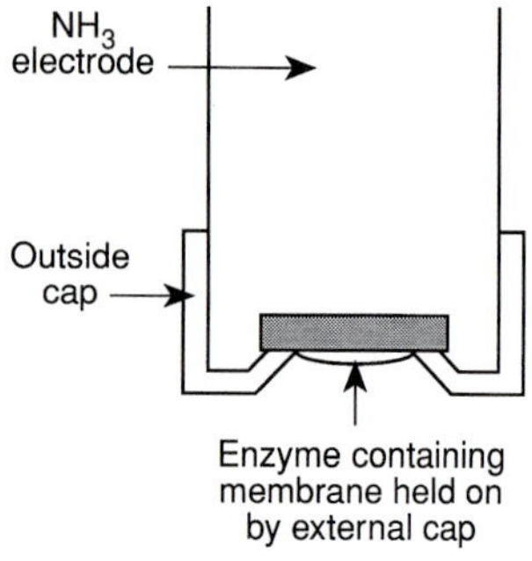

Fig. 5.7 The modification of a NH_3 electrode for use as an enzyme electrode to detect, for example, nitrite or urea.

The previous section showed how molecules like CO_2 and NH_3 can be determined by potentiometric sensors. These in turn can be deployed to detect more complex species, not themselves amenable to direct potentiometric analysis, by using them in conjunction with a membrane in which a suitable enzyme is immobilised. Figure 5.7 illustrates the concept in the specific context of an ammonia electrode. The purpose of the enzyme is to convert the substrate to be detected into NH_3 which is then detected by the ammonia sensor to give a quantitative measure of the substrate concentration. Typical substrates and enzymes are

- nitrite, NO_2^-, and nitrite reductase
- urea, NH_2CONH_2, and urease

Changing the NH_3 electrode appropriately to a different potentiometric sensor enables a wide range of species to be detected, ranging from glucose to amino-acids.

5.6 Ion-selective field effect transistors (ISFETs)

Earlier sections of this chapter have illustrated potentiometric methods for the selective determination of ions and molecules. An alternative, elegant, more recent adaptation in which the currents flowing in tiny field effect transistors (FETs) can be used to probe ion concentrations is shown schematically in Fig. 5.8. The FET is made of n-p-n semiconductors which form respectively the *source, channel* and the *drain*. In a conventional FET the current flowing between the source and the drain is influenced by a voltage applied to the (metallic) *gate* (Fig. 5.8). In an ISFET an ion-selective membrane acts as the gate instead of a metal and the electrical potential it exerts on the source-to-drain current depends on the extent to which substrate ions have entered the membrane. The current therefore measures the ionic concentrations. Typical ion-membrane combinations for use in ISFETs are

- H^+ and a hydrated silica membrane
- Cl^- and a AgCl membrane
- K^+ and a valinomycin containing membrane
- I^- and CN^- and either an AgI or AgCN membrane.

In this way selective determinations of ion concentrations can be conducted on the surface of a tiny silicon chip.

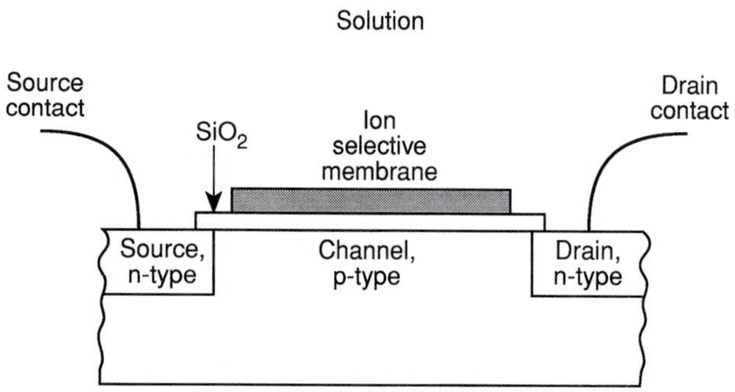

Fig. 5.8 An ion-selective field effect transistor.

6 Worked examples and problems

In this chapter some problems based on standard electrode potentials are provided for the reader to test his understanding. As a preliminary a number of worked examples are given.

6.1 Worked examples

(1) Equilibrium constants

Calculate the equilibrium constants for the following reactions at 25°C in aqueous solutions,

$$(a)\ Sn(s) + CuSO_4(aq) \rightleftharpoons Cu(s) + SnSO_4(aq)$$

$$(b)\ 2H_2(g) + O_2(g) \rightleftharpoons 2H_2O(l)$$

given the following standard electrode potentials

$$\tfrac{1}{2}Sn^{2+}(aq) + e^- \rightleftharpoons \tfrac{1}{2}Sn(s) \qquad -0.136\ V$$

$$\tfrac{1}{2}Cu^{2+}(aq) + e^- \rightleftharpoons \tfrac{1}{2}Cu(s) \qquad +0.337\ V$$

$$\tfrac{1}{4}O_2(g) + H^+(aq) + e^- \rightleftharpoons \tfrac{1}{2}H_2O(l) \qquad +1.229\ V$$

We consider first equilibrium (*a*) and begin by noting that definition of the standard electrode potential of the Sn/Sn^{2+} couple implies that for the following cell,

As a general strategy for problem solving as a start always write down a cell diagram and work out the associated (formal) cell reaction.

$$\text{Pt} \mid \text{H}_2\ (\text{g})\ (P = 1\ \text{atm}) \mid \text{H}^+(\text{aq})\ (a = 1) \parallel \text{Sn}^{2+}(\text{aq})\ (a = 1) \mid \text{Sn}$$

the cell potential is,

$$E^\ominus_{\text{Sn/Sn}^{2+}} = -0.136\ \text{V} = \phi_{\text{Sn}} - \phi_{\text{Pt}}$$

The strategy commended in Section (1.13) allows us to associate a *formal* cell reaction with the above cell as follows. The potential determining equilibrium at the right hand electrode is:

$$\tfrac{1}{2}\text{Sn}^{2+}(\text{aq}) + \text{e}^- \rightleftharpoons \tfrac{1}{2}\text{Sn(s)}$$

and at the left hand electrode:

$$\text{H}^+(\text{aq}) + \text{e}^- \rightleftharpoons \tfrac{1}{2}\text{H}_2(\text{g})$$

Subtracting gives

$$\tfrac{1}{2}\text{Sn}^{2+}(\text{aq}) + \tfrac{1}{2}\text{H}_2(\text{g}) \rightleftharpoons \tfrac{1}{2}\text{Sn(s)} + \text{H}^+(\text{aq}) \qquad (6.1)$$

For this last reaction

$$\Delta G^\ominus = -FE^\ominus_{\text{Sn/Sn}^{2+}} = +0.136F$$

Likewise for the cell

$$\text{Pt} \mid \text{H}_2\ (\text{g})\ (P = 1\ \text{atm}) \mid \text{H}^+(\text{aq})\ (a = 1) \parallel \text{Cu}^{2+}(\text{aq})\ (a = 1) \mid \text{Cu}$$

the cell potential is,

$$E^\ominus_{\text{Cu/Cu}^{2+}} = +0.337\ \text{V}$$

The potential determining equilibria at each electrode are

Right Hand Electrode: $\frac{1}{2}Cu^{2+}(aq) + e^- \rightleftharpoons \frac{1}{2}Cu(s)$

Left Hand Electrode: $H^+(aq) + e^- \rightleftharpoons \frac{1}{2}H_2(g)$

This enables the formal cell reaction to be deduced (Section 1.13):

$$\tfrac{1}{2}Cu^{2+}(aq) + \tfrac{1}{2}H_2(g) \rightleftharpoons \tfrac{1}{2}Cu(s) + H^+(aq) \qquad (6.2)$$

for which

$$\Delta G^\ominus = -FE^\ominus_{Cu/Cu^{2+}} = -0.337F$$

From reactions (6.1) and (6.2), subtracting,

$$\tfrac{1}{2}Cu^{2+}(aq) + \tfrac{1}{2}Sn(s) \rightleftharpoons \tfrac{1}{2}Cu(s) + \tfrac{1}{2}Sn^{2+}(aq) \qquad (6.3)$$

for which,

Neither Cu nor Sn feature in the definition of K_c since they are pure solids as explained in Box 1.8.

$$\begin{aligned}\Delta G^\ominus &= (-0.337F) - (+0.136F) \\ &= -0.473F \\ &= -RT \ln K_c\end{aligned}$$

where,

It is only meaningful to cite a value of an equilibrium constant (or value of $\Delta G^\ominus$, $\Delta H^\ominus$, etc) if the associated chemical equation is written down. In the present case this is reaction (6.3).

$$\begin{aligned}K_c &= \frac{[Sn^{2+}]^{\frac{1}{2}}}{[Cu^{2+}]^{\frac{1}{2}}} \\ &= \exp\left(\frac{0.473F}{RT}\right) \\ &= 1 \times 10^8\end{aligned}$$

Returning to the original question we note that for

$$Sn(s) + CuSO_4(aq) \rightleftharpoons Cu(s) + SnSO_4(aq)$$

the equilibrium constant,

$$\frac{[Sn^{2+}]}{[Cu^{2+}]} = (10^8)^2 = 10^{16}$$

Next we turn to equilibrium (*b*). The last standard electrode potential cited in the problem relates to the following cell:

$Pt \mid H_2\,(g)\,(P = 1\text{ atm}) \mid H^+(aq)\,(a = 1), H_2O(a = 1) \mid O_2(g)(P = 1\text{ atm}) \mid Pt$

for which the formal cell reaction can be deduced by subtracting the potential determining equilibria at each electrode, as follows

Right Hand Electrode : $\frac{1}{4}O_2(g) + H^+(aq) + e^- \rightleftharpoons \frac{1}{2}H_2O(l)$

Left Hand Electrode : $H^+(aq) + e^- \rightleftharpoons \frac{1}{2}H_2(g)$

to give

$$\tfrac{1}{4}O_2(g) + \tfrac{1}{2}H_2(g) \rightleftharpoons \tfrac{1}{2}H_2O(l)$$

For this reaction,

$$\Delta G^\ominus = -1.229F$$

and the associated equilibrium constant

$$K = \frac{1}{P_{H_2}^{\frac{1}{2}} P_{O_2}^{\frac{1}{4}}} = \exp\left(-\frac{\Delta G^\ominus}{RT}\right)$$

The interpretation of $\Delta G^\ominus$ is through the equation
$\Delta G^\ominus = \frac{1}{2}\mu^\nabla_{H_2O(l)} - \frac{1}{2}\mu^o_{H_2(g)} - \frac{1}{4}\mu^o_{O_2(g)}$
so that
$\Delta G^\ominus$ = {Gibbs free energy of 0.5 moles of pure liquid H_2O}
−{Gibbs free energy of 0.5 moles of pure H_2 gas at 1 atmosphere pressure}
−{Gibbs free energy of 0.25 moles of pure O_2 gas at 1 atmosphere pressure}
where all the free energies are measured at the standard temperature of 298 K.

where P_{H_2} and P_{O_2} are the partial pressures of hydrogen and oxygen, is given by

$$K = \exp\left(\frac{1.229F}{RT}\right) = 6 \times 10^{20} (\text{atmosphere})^{-\frac{3}{4}}$$

Notice that the activity of water is absent from the definition of K for the reasons explained in Box 1.8.

(2) The Nernst equation

For the following cell

$$Al \mid Al^{3+}\ (aq) \parallel Sn^{4+}(aq),\ Sn^{2+}(aq) \mid Pt$$

state or calculate at 25°C, (a) the cell reaction; (b) the cell EMF when all concentrations are 0.1 M and 1.0 M (ignore activity coecients); (c) ΔG° for the cell reaction in (a); (d) K for the cell reaction in (a); (e) the positive electrode and the direction of electron flow in an external circuit connecting the two electrodes. The standard electrode potentials are $E^{\circ}_{Sn^{2+}/Sn^{4+}} = 0.15$ V and $E^{\circ}_{Al/Al^{3+}} = -1.61$ V.

The potential determining equilibria are

Right Hand Electrode : $\frac{1}{2}Sn^{4+}(aq) + e^-(metal) \rightleftharpoons \frac{1}{2}Sn^{2+}(aq)$

Left Hand Electrode : $\frac{1}{3}Al^{3+}(aq) + e^-(metal) \rightleftharpoons \frac{1}{3}Al(s)$

So that the formal cell reaction is

$$\tfrac{1}{2}Sn^{4+}(aq) + \tfrac{1}{3}Al(s) \rightleftharpoons \tfrac{1}{2}Sn^{2+}(aq) + \tfrac{1}{3}Al^{3+}(aq) \qquad (6.4)$$

When all the potential determining species in the cell are present at unit activity the cell potential is

$$\begin{aligned} E^{\circ}_{cell} &= E^{\circ}_{Sn^{2+}/Sn^{4+}} - E^{\circ}_{Al/Al^{3+}} \\ &= (0.15) - (-1.66) \\ &= 1.81 \text{ V} \end{aligned}$$

so that for reaction (6.4),

$$\begin{aligned} \Delta G^{\circ} &= -1.81F \\ &= -175 \text{ kJ mol}^{-1} \end{aligned}$$

It follows that reaction (6.4) is thermodynamically downhill and is the process which would occur if the cell was short circuited.

Notice in this example the *formal* cell reaction is the same as the *spontaneous* cell reaction (see discussion in Section 1.13).

Following Section 1.13 the cell EMF will be given by the appropriate Nernst equation,

$$E = 1.81 - \frac{RT}{F}\ln\left\{\frac{[Al^{3+}]^{\frac{1}{3}}[Sn^{2+}]^{\frac{1}{2}}}{[Sn^{4+}]^{\frac{1}{2}}}\right\}$$

So that when all the concentrations are 1.0 M the cell EMF is 1.81 V. When the concentrations are 0.1 M,

$$\begin{aligned} E &= 1.81 - \left(\frac{RT}{F}\right)\ln\left\{\frac{0.1^{\frac{1}{3}}0.1^{\frac{1}{2}}}{0.1^{\frac{1}{2}}}\right\} \\ &= 1.81 + 0.02 \\ &= 1.83 \text{ V} \end{aligned}$$

The equilibrium constant for the reaction is

$$K = \frac{[Al^{3+}]^{\frac{1}{3}}[Sn^{2+}]^{\frac{1}{2}}}{[Sn^{4+}]^{\frac{1}{2}}} = \exp\left(\frac{1.81F}{RT}\right)$$

$$= 4 \times 10^{30}(\text{moles dm}^{-3})^{\frac{1}{3}}$$

Section 1.12 discusses the sign convention for reporting cell polarities.

Last we note the cell polarity will always be

$$\ominus \text{ Al} \mid Al^{3+}\text{ (aq)} \parallel Sn^{4+}\text{(aq)}, Sn^{2+}\text{(aq)} \mid \text{Pt} \oplus$$

unless tremendous extremes of concentration ratios occur ($[Al^{3+}] >> [Sn^{4+}]$ and $[Sn^{2+}] >> [Sn^{4+}]$). It follows that if the cell is short circuited then electrons would leave the aluminium, which would oxidise

$$\tfrac{1}{3}\text{Al(s)} \rightarrow \tfrac{1}{3}Al^{3+}\text{(aq)} + e^-$$

and that at the platinum electrode Sn^{4+} ions would be reduced

$$\tfrac{1}{2}Sn^{4+}\text{(aq)} + e^- \rightarrow \tfrac{1}{2}Sn^{2+}\text{(aq)}$$

This inference is consistent with the nature of the spontaneous cell reaction deduced above.

(3) Concentration cells

Consider the following cell,

$$Pt \mid H_2\,(g)\;(P_1) \mid HCl\,(aq)\;(m_1\ \text{M}) \parallel HCl\,(aq)\;(m_2\ \text{M}) \mid H_2\,(g)\;(P_2) \mid Pt$$

where the hydrogen gas pressures are P_1 and P_2 atmospheres respectively and the two hydrochloric acid concentrations, m_1 and m_2 (moles dm^{-3}). At 25°C calculate or state (a) an expression for the cell EMF in terms of m_1, m_2, P_1 and P_2 (ignoring activity coefficients); (b) the cell EMF when $m_1 = 0.1$ M, $m_2 = 0.2$ M and $P_1 = P_2 = 1$ atm; (c) the cell EMF when the hydrogen pressure P_2 is increased to 10 atm, all other concentrations remaining the same; (d) the cell reaction.

As in the previous examples the strategy is first to identify the potential determining equilibria which in this case are,

Right Hand Electrode : $H^+(\text{aq}, m_2) + e^-(\text{metal}) \rightleftharpoons \tfrac{1}{2}H_2(\text{g}, P_2)$

Left Hand Electrode : $H^+(\text{aq}, m_1) + e^-(\text{metal}) \rightleftharpoons \tfrac{1}{2}H_2(\text{g}, P_1)$

Second the *formal* cell reaction is (Section 1.13):

$$H^+(\text{aq}, m_2) + \tfrac{1}{2}H_2(\text{g}, P_1) \rightleftharpoons H^+(\text{aq}, m_1) + \tfrac{1}{2}H_2(\text{g}, P_2)$$

The Nernst equation is therefore

$$E = 0.0 - \left(\frac{RT}{F}\right)\ln\left\{\frac{P_2^{\frac{1}{2}}\, m_1}{m_2\, P_1^{\frac{1}{2}}}\right\}$$

When $P_1 = P_2 = 1$ and $m_1 = 0.1$ M but $m_2 = 0.2$ M,

$$E = -\left(\frac{RT}{F}\right)\ln\left\{\frac{0.1}{0.2}\right\}$$

$$= 0.018 \text{ V}$$

If P_2 is changed to 10 atm we then have,

$$E = -\left(\frac{RT}{F}\right)\ln\left\{\frac{0.1 \times 10^{\frac{1}{2}}}{0.2}\right\}$$
$$= -0.012 \text{ V}$$

The *formal* cell reaction in this example was established above, The spontaneous cell reaction—that occurring when the cell is short circuited—can be seen from the above to depend on the cell concentrations. When $P_1 = P_2 = 1, m_1 = 0.1 \text{ M} = 0.5\, m_2$ then the spontaneous reaction is the same as the formal cell reaction since

The potentials generated in concentration cells are tiny, typically of the order of tens of millivolts , as in this example.

$$\Delta G = -0.018F < 0$$

However when P_2 is increased to 10 atmospheres,

$$\Delta G = +0.012F > 0$$

so that the direction of the spontaneous cell reaction is the opposite of the formal cell reaction (see Section 1.13).

(4) Solubility Products

Given the standard electrode potentials,

$$E^{\ominus}_{Ag/Ag^+} = +0.799 \text{ V and } E^{\ominus}_{Ag/AgI} = -0.152 \text{ V},$$

calculate the solubility product, K_{sp}, and solubility of silver iodide at 25°C.

The first standard electrode potential quoted relates to the cell,

$$\text{Pt} \mid H_2\,(g)(P = 1 \text{ atm}) \mid H^+(aq)\,(a = 1) \parallel Ag^+(aq)\,(a = 1) \mid Ag$$

for which the formal cell reaction is readily established as,

$$Ag^+\,(aq) + \tfrac{1}{2}H_2\,(g) \rightleftharpoons Ag\,(s) + H^+\,(aq) \qquad (6.5)$$

and for which,

$$\Delta G^{\ominus} = -0.799F$$

Likewise the second standard electrode potential is that of the cell,

$$\text{Pt} \mid H_2\,(g)(P = 1 \text{ atm}) \mid H^+(aq)\,(a = 1), I^-(aq)\,(a = 1) \mid AgI \mid Ag$$

The potential determining equilibria at the two electrodes are

Right Hand Electrode : $AgI(s) + e^-(metal) \rightleftharpoons Ag(s) + I^-(aq)$

Left Hand Electrode : $H^+(aq) + e^-(metal) \rightleftharpoons \tfrac{1}{2}H_2(g)$

so that the formal cell reactions is

$$AgI(s) + \tfrac{1}{2}H_2(g) \rightleftharpoons Ag(s) + I^-(aq) + H^+(aq) \qquad (6.6)$$

which has,

$$\Delta G^{\ominus} = +0.152F$$

Subtracting reactions (6.6) and (6.5) gives

$$AgI(s) \rightleftharpoons Ag^+(aq) + I^-(aq)$$

for which,

$$\Delta G^{\ominus} = (+0.152F) - (-0.799F)$$
$$= +0.951F$$
$$= -RT\ln K_{sp}$$

Hence

$$K_{sp} = \exp\left\{\frac{-0.951F}{RT}\right\} = [Ag^+][I^-]$$
$$= 8.5 \times 10^{-17} mol^2 dm^{-6}$$

This corresponds to solubility of 2.2×10^{-6}g dm^{-3} at 25°C.

(5) Weak acids

The EMF of each of the following Harned cells is measured at two temperatures

$$Pt \mid H_2\,(g)\ (P = 1\ atm) \mid HCl\ (10^{-5}\ M) \mid AgCl \mid Ag\ (E_1)$$

$$Pt|H_2\,(g)\ (P = 1)|HA\ (10^{-2}\ \text{M}),\ KA\ (10^{-2}\ \text{M}),\ KCl\ (10^{-5}\ \text{M})|AgCl|Ag\ (E_2)$$

where HA is a weak acid and KA is its potassium salt. The results are as follows:

	293 K	*303 K*
$E_1 V$	*0.820*	*0.806*
$E_2 V$	*0.878*	*0.866*

Calculate K_a and $\Delta H^{\ominus}$ for the dissociation of the weak acid, pointing out any assumptions you make. Do not ignore activity coefficients, but assume in the second cell that $[HA] >> [H^+]$.

We start by identifying the potential determining equilibria in the two cells. In *both* cases these are

Right Hand Electrode : $AgCl(s) + e^-(metal) \rightleftharpoons Ag(s) + Cl^-(aq)$
Left Hand Electrode : $H^+(aq) + e^-(metal) \rightleftharpoons \frac{1}{2}H_2(g)$

so that the formal cell reaction is

$$AgCl(s) + \tfrac{1}{2}H_2(g) \rightleftharpoons Ag(s) + Cl^-(aq) + H^+(aq)$$

The corresponding Nernst equation is

$$E = E^{\ominus} - (RT/F) \ln \{a_{H^+}.a_{Cl^-}\}$$

The absence of a_{Ag} and a_{AgCl} from the Nernst equation is explained in Section 1.7 on the Ag/AgCl electrode. Note also that $P_{H_2} = 1$ atm.

where a_{H^+} and a_{Cl^-} are the activities of the H^+ and Cl^- ions. This equation applies to *both* Harned cells although the values of a_{H^+} and a_{Cl^-} differ between them. Now

$$a_{H^+} = \gamma_{H^+}[H^+] \text{ and } a_{Cl^-} = \gamma_{Cl^-}[Cl^-]$$

where γ is the appropriate activity coefficient.

Consider the first cell and apply the Nernst equation at the lower temperature (293 K):

$$0.820 = E^{\ominus}_{Ag/AgCl} - (293R/F) \ln\{\gamma_{H^+}.\gamma_{Cl^-}.10^{-5}.10^{-5}\}$$

or

$$0.820 = E^{\ominus}_{Ag/AgCl} - 0.058 \log_{10}\{\gamma_{H^+}.\gamma_{Cl^-}.10^{-10}\}$$

However at concentrations as low as 10^{-5} M the solutions are effectively ideal to a high degree of approximation. Physically this arises since the ions are so far apart that the ion–ion interactions (Chapter 2) are negligible. We can therefore put

$$\gamma_{H^+} \approx \gamma_{Cl^-} \approx 1$$

and so deduce that

$$E^{\ominus}_{Ag/AgCl} = 0.240 \text{ V (at 293 K)}$$

Considering the data for the higher temperature,

$$0.806 = E^{\ominus}_{Ag/AgCl} - (303R/F)\ \ln\{\gamma_{H^+}.\gamma_{Cl^-}.10^{-5}.10^{-5}\}$$

or

$$0.820 = E^{\ominus}_{Ag/AgCl} - 0.060\ \log_{10}\{\gamma_{H^+}.\gamma_{Cl^-}.10^{-10}\}$$

so that

$$E^{\ominus}_{Ag/AgCl} = 0.206 \text{ V (at 303 K)}$$

We next turn to the second cell and note that the hydrogen ion activity, a_{H^+}, 'seen' by the hydrogen electrode will be governed by the dissociation of the weak acid,

$$HA(aq) \rightleftharpoons H^+(aq) + A^-(aq)$$

for which we can write the acid dissociation constant,

$$K_a = \frac{a_{H^+} \cdot a_{A^-}}{a_{HA}} = \frac{\gamma_{H^+}\gamma_{A^-}}{\gamma_{HA}} \cdot \frac{[H^+][A^-]}{[HA]}$$

Now HA is uncharged so that we can safely assume,

$$\gamma_{HA} \approx 1$$

is a very good approximation. However the ionic strength, I, of the solution is in excess of 10^{-2} mol dm^{-3} so, following Chapter 2, we expect that

$$\gamma_{Cl^-} < 1 \text{ and } \gamma_{A^-} < 1$$

Returning to the Nernst equation,

$$E_2 = E^{\ominus}_{Ag/AgCl} - (RT/F)\ \ln\{a_{H^+}.\gamma_{Cl^-}.10^{-5}\}$$

At the lower temperature of 293 K

$$0.878 = 0.240 - 0.058\log_{10}\{a_{H^+} \cdot \gamma_{Cl^-}.10^{-5}\}$$

so that

$$\log_{10}\{a_{H^+}.\gamma_{Cl^-}.\} = -6.00 \text{ (at 293 K)}$$

At the higher temperature,

$$0.866 = 0.206 - 0.060\log_{10}\{a_{H^+}.\gamma_{Cl^-}.10^{-5}\}$$

giving

$$\log_{10}\{a_{H^+}.\gamma_{Cl^-}.\} = -6.00 \text{ (at 303 K)}$$

It follow that at both temperatures

$$a_{H^+}.\gamma_{Cl^-}. = 10^{-6}$$

Now,

$$K_a = a_{H^+}.\gamma_{Cl^-}\left(\frac{\gamma_{A^-}}{\gamma_{Cl^-}}\right).\frac{[A^-]}{[HA]}$$

$$= 10^{-6}.\left(\frac{\gamma_{A^-}}{\gamma_{Cl^-}}\right).\frac{10^{-2}}{10^{-2}}$$

$$= 10^{-6}\text{M}$$

if $\gamma_{A^-} = \gamma_{Cl^-}$. This last assumption is good since the Debye–Hückel limiting law,

$$\log_{10}\gamma = -Az^2\sqrt{I}$$

predicts the same value for the activity coefficients of ions with the same charge, z, experiencing the same ionic strength, I.

The constant A depends only on the temperature and nature of the solvent.

We have shown that K_a has the same value (10^{-6} M) at both 293 K and 303 K. We can find ΔH for the acid dissociation by using the van't Hoff isochore,

$$\frac{d\ln K}{dT} = \frac{\Delta H^\ominus}{RT^2}$$

which shows that

$$\Delta H^\ominus \approx 0$$

Since HA is a weak acid, $K_a << 1$ and $\Delta G_a >> 0$. If $\Delta H^\ominus \approx 0$ *it follows that* $\Delta S^\ominus << 0$ and the *weakness* of the acid arises from entropic not enthalpic effects. Such behaviour is shown by carboxylic acids, such as CH_3COOH. Their *weak* acid behaviour is caused by the unfavourable entropy of dissociation which arises from ordering of the solvent molecules by the charged dissociation products, H^+ and $RCOO^-$.

(6) Thermodynamic quantities

The EMF of the cell

$$Ag \mid AgCl \mid HCl\ (10^{-5}\ \text{M}) \mid Hg_2Cl_2 \mid Hg$$

is 0.0421 V at 288 K and 0.0489 V at 308 K. Use this infomation to calculate the enthalpy, free energy and entropy changes that accompany the cell reaction at 298 K.

The potential determining equilibria at the electrodes are

Right Hand Electrode: $\frac{1}{2}Hg_2Cl_2(s) + e^-(\text{metal}) \rightleftharpoons Hg(l) + Cl^-(aq)$

Left Hand Electrode: $AgCl(s) + e^-(\text{metal}) \rightleftharpoons Ag(s) + Cl^-(aq)$

On subtracting the following formal cell reaction is generated,

$$\tfrac{1}{2}Hg_2Cl_2(s) + Ag(s) \rightleftharpoons Hg(l) + AgCl(s) \qquad (6.7)$$

for which

$$\Delta G^\ominus = -FE^\ominus$$

so

$$\Delta G^\ominus_{288} = -0.0421F$$
$$= -4.062 \text{ kJ mole}^{-1} \quad \text{at 288 K}$$

while

$$\Delta G^\ominus_{308} = -0.0489F$$
$$= -4.719 \text{ kJ mole}^{-1} \text{ at 308 K}$$

Linearly interpolating between 288 K and 308 K we find

$$\Delta G^\ominus_{298} = -4.390 \text{ kJ mole}^{-1} \quad \text{at 298 K}$$

Notice that the reaction (6.7) involves the pure solid metal chlorides and pure elements in their standard states so that the free energies evaluated above are *standard* free energies regardless of the concentration of HCl in the cell—the latter does not enter the *net* formal cell reaction, or influence the cell EMF. It does however play the vital role of establishing the potentials on the two electrodes through the potential determining equilibria given above.

The entropy change can be found from

$$\Delta S^{\ominus}_{298} = F\left(\frac{\partial E^{\ominus}}{\partial T}\right)_P$$

$$= F\left\{\frac{0.0489 - 0.0421}{20}\right\}$$

$$= 32.8 \text{ JK}^{-1} \text{ mole}^{-1} \quad \text{at 298 K}$$

Then the enthalpy change may be readily estimated as follows:

$$\Delta H^{\ominus} = \Delta G^{\ominus} + T\Delta S^{\ominus}$$

$$\Delta H^{\ominus}_{298}/\text{kJ mole}^{-1} = -4.390 + \left\{\frac{298 \times 32.8}{10^3}\right\}$$

$$= +5.387 \quad \text{at 298 K}$$

It is apparent that the cell reaction is thermodynamically downhill ($\Delta G < 0$) but that it is *entropy driven* ($\Delta S > 0$), the process being enthalphically unfavourable ($\Delta H > 0$). The positive ΔS value reflects the increase in disorder in converting the solids Hg_2Cl_2 and Ag into solid AgCl and *liquid* Hg.

(7) Activity coefficients

Note that Box 4.1 gives a worked example showing how EMF data may be used to quantify activity coefficients.

6.2 Problems

$R = 8.313$ J K^{-1} mole^{-1}; $F = 96490$ C mole^{-1}

(1) Consider the following cell at 298 K:

$$\text{Zn} \mid \text{Zn}^{2+}(\text{aq})\ (a = 0.1) \parallel \text{Cu}^{2+}(\text{aq})\ (a = 1) \mid \text{Cu}$$

Calculate the cell EMF given that $E^{\ominus}_{\text{Cu/Cu}^{2+}} = +0.34$ V and $E^{\ominus}_{\text{Zn/Zn}^{2+}} = -0.76$ at 298 K.

(2) The EMF of the cell

$$\text{Pb (s)} \mid \text{Pb}^{2+}(\text{aq})\ (a = 1) \parallel \text{Ag}^{+}(\text{aq})\ (a = 0.5) \mid \text{Ag (s)}$$

is 0.912 V at 25°C. Write down the cell reaction and, given that the standard electrode potential of the Pb/Pb^{2+} electrode is –0.130 V, determine the standard electrode potential of the Ag/Ag^{+} electrode.

(3) At 25°C the cell,

$$\text{Pt} \mid \text{Fe}^{2+}(\text{aq})\ (a = 1), \text{Fe}^{3+}(\text{aq})\ (a = 1) \parallel \text{Ce}^{3+}(\text{aq})\ (a = 1), \text{Ce}^{4+}(\text{aq})\ (a = 1) \mid \text{Pt}$$

has an EMF of +0.84 V. Calculate the equilibrium constant of the cell reaction.

(4) The EMF of the cell

$$\text{Pt} \mid \text{H}_2\ (\text{g})\ (P = 1 \text{ atm}) \mid \text{HBr}\ (10^{-4}\ \text{M}) \mid \text{CuBr(s)} \mid \text{Cu}$$

is 0.559 at 25°C. The standard electrode potential for the Cu/Cu^{+} couple is 0.522 V. Calculate the solubility product of CuBr.

(5) The standard electrode potential of the cell

$$\mathrm{Pt} \mid \mathrm{H_2}\ (\mathrm{g})\ (P = 1\ \mathrm{atm}) \mid \mathrm{HCl}\ (a = 1) \mid \mathrm{AgCl}\ (\mathrm{s}) \mid \mathrm{Ag}$$

can be expressed in the form

$$E^{\ominus}/\mathrm{V} = 0.2366 - 4.856 \times 10^{-4}(T-273) - 3.421 \times 10^{-6}(T-273)^2$$

where T is the temperature measured in K.

Write down the cell reaction and calculate ΔG, ΔS and ΔH at 298 K.

(6) Solid AgCl conducts electricity sufficiently that the cell

$$\mathrm{Ag} \mid \mathrm{AgCl} \mid \mathrm{Cl_2}\ (\mathrm{g}) \mid \mathrm{Pt}$$

is reversible with the AgCl either solid or liquid. The EMF of the cell as a function of temperature is given below

T/K	573	623	673	723	773	823	873
E/V	1.000	0.975	0.949	0.924	0.904	0.887	0.871

Calculate the enthalpy and entropy of fusion and the melting point of AgCl.

(7) Write down the cell reaction for the Harned cell,

$$\mathrm{Pt} \mid \mathrm{H_2}\ (\mathrm{g})\ (P = 1\ \mathrm{atm}) \mid \mathrm{HCl}\ (\mathrm{aq}) \mid \mathrm{AgCl}\ (\mathrm{s}) \mid \mathrm{Ag}$$

and obtain an expression for the EMF, E, assuming that the gas phase species behaves ideally.

The data given below give the EMF for the Harned cell at 25°C and as a function of the molality, m, of the $\mathrm{Cl^-}$ ion. Using the Debye–Hückel limiting law, determine the standard electrode potential of the silver/silver chloride electrode.

E/V	m/mol kg^{-1}
0.62565	4×10^{-4}
0.58460	9×10^{-4}
0.55565	1.6×10^{-3}
0.53312	2.5×10^{-3}

(8) In a study of the two separate cells,
Cell A,

$$\mathrm{Pt} \mid \mathrm{H_2}\ (\mathrm{g})\ (P = 1\ \mathrm{atm}) \mid \mathrm{NaOH}\ (\mathrm{aq})(c\ \mathrm{M}),\ \mathrm{NaCl}\ (\mathrm{aq})(c\ \mathrm{M}) \mid \mathrm{AgCl}\ (\mathrm{s}) \mid \mathrm{Ag}$$

and Cell B,

$$\mathrm{Pt} \mid \mathrm{H_2}\ (\mathrm{g})\ (P = 1\ \mathrm{atm}) \mid \mathrm{HCl}\ (\mathrm{aq})(c\ \mathrm{M}),\ \mathrm{NaCl}\ (\mathrm{aq})(c\ \mathrm{M}) \mid \mathrm{AgCl}\ (\mathrm{s}) \mid \mathrm{Ag}$$

the EMFs, E, were measured as a function of the concentrations of electrolytes at 25°C:

c/mol dm^{-3}	E(Cell A)/V	E(Cell B)/V
0.001	1.0493	0.5780
0.005	1.0493	0.4969
0.010	1.0493	0.4624

Determine a value for the dissociation constant, K_w, of water, where

$$K_w = a_{H^+} \cdot a_{Cl^-}$$

(9) EMF measurements were made at 25°C on the cell

Pt | H_2 (g) ($P = 1$ atm) | NH_3 (aq)(c M), NH_4Cl (aq)(c M) | AgCl (s) | Ag

as follows:

c/mol dm^{-3}	0.00857	0.0158	0.0212	0.0330
E/V	0.8965	0.8824	0.8759	0.8661

Given that the standard electrode potential for the silver/silver chloride electrode at 25°C is 0.2223 V, determine the acid dissociation constant for NH_4^+. Assume that the activity coefficient of NH_3 is unity.

$\Delta H^{\ominus}_{298}$ for the reaction

$$NH_3(aq) + H_2O(l) \rightleftharpoons NH_4^+(aq) + OH^-(aq)$$

is 4.36 kJ mol^{-1}. Calculate $\Delta S^{\ominus}_{298}$ using the fact that the dissociation constant of water, K_w, is 10^{-14} mol^2 dm^{-6} at 25°C.

(10) At 298 K and at one atmosphere pressure, the EMF, E, of the cell,

Cd (amalgam)(4.6% Cd) | $CdCl_2$ (aq) (c M) | AgCl (s) | Ag

varies with the concentration, c, of $CdCl_2$ as follows

$c/10^{-3}$mol dm^{-3}	0.1087	0.1269	0.2144	0.3659
E/V	0.9023	0.8978	0.8803	0.8641

Write down the Nernst equation for this cell and, using a Debye–Hückel extrapolation procedure, deduce the cell EMF when both Cd^{2+} and Cl^- are at unit activity in the solution.

Under the same conditions, the EMF of the cell

Cd (amalgam)(4.6% Cd) | $CdCl_2$ (aq) (0.5 M) | Cd

is −0.0534 *V*. If the standard electrode potential of the silver/silver chloride electrode is 0.2222 V, (*a*) What is the standard electrode potential of the Cd/Cd^{2+} couple? (*b*) What is the Gibbs free energy of formation, $\Delta G_f^{\ominus}$, of Cd^{2+}(aq)? (*c*) What is the mean ionic activity coefficient of 0.2144×10^{-3} M $CdCl_2$?